Analog Circuits and Signal Processing

For further volumes:
http://www.springer.com/series/7381

Hagen Marien · Michiel Steyaert
Paul Heremans

Analog Organic Electronics

Building Blocks for Organic Smart Sensor Systems on Foil

Springer

Hagen Marien
Dept. Elektrotechniek
K.U. Leuven
Kardinaal Mercierlaan 94
B-3001 Heverlee
Belgium

Paul Heremans
Interuniversity Microelectronics
    Centre (imec)
Kapeldreef 75
3001 Leuven
Belgium

Michiel Steyaert
Dept. Elektrotechniek
K.U. Leuven
Kardinaal Mercierlaan 94
B-3001 Heverlee
Belgium

ISBN 978-1-4899-8547-7   ISBN 978-1-4614-3421-4 (eBook)
DOI 10.1007/978-1-4614-3421-4
Springer New York Heidelberg Dordrecht London

Printed on acid-free paper

Springer is part of Springer Science+Business Media (www.springer.com)

Preface

Today's trend in mobile electronics toward smaller and lighter devices conflicts with the other market-driven interest for larger displays. Large displays can improve the user-friendliness of the devices. By using flexible and rollable displays made in organic electronics technology, small devices with large displays are enabled. In this technology a flexible plastic foil is used as the substrate for the electronic circuits. As a result, the production of devices with flexible or even rollable displays is enabled. Besides the opportunities for flexible displays, there is also an interest for other applications, such as RFID, printed solar cells, flexible lighting and organic smart sensor systems. For the realization of these applications mostly digital building blocks are required. However, most of them also need analog building blocks for sensors, sensor read-out, and analog-to-digital conversion. The design of analog organic circuits is the subject of this research work.

The main advantages of organic electronics technology are the low production temperatures, the low-cost, and flexible plastic substrate, and the printability. The production temperature for organic transistors stays below 150 °C during the process. This is very low compared to temperatures up to 1000 °C in standard Si CMOS technology. The low temperatures enable the use of low-cost flexible materials, such as a plastic foil, for substrates in the technology. Those materials would melt or burn at 1000 °C. Furthermore, in recent years the solution-based deposition and the printing of organic electronics circuits have been reported. The combination of all these advantages create a niche domain of large-area, large-scale flexible applications.

Despite all the advantageous properties, at the moment organic electronics technology suffers from some adverse properties that hamper the design of both digital and analog organic circuits. The intrinsic gain of organic transistors is very low (typically ~ 5). Only a limited component library is available at present. As active elements, only p-type transistors and, as passive elements, only capacitors are available. Resistive behavior is found through transistors biased in the linear regime. Furthermore, organic technology suffers from the variation of behavioral transistor parameters which result in mismatch and reliability issues for analog circuits.

In this work the properties of organic electronics technology have been observed first. Then the design strategy for analog circuits has been adapted in such a way that adverse properties become harmless and that advantageous properties are fully exploited. One property of the organic transistors is the availability of a backgate pin. The use of this backgate pin is investigated and multiple techniques have been proposed to fully exploit this pin, e.g., for tuning the threshold voltage of the transistor.

In this dissertation the design of several analog building blocks for organic smart sensor systems are presented. The presented building blocks are amplifiers, analog-to-digital converters, sensors, and DC-DC converters.

A single-stage differential amplifier is presented. Several analog techniques are applied to optimize the reliability and the gain of the single-stage amplifier. Bootstrapped gain enhancement is applied to the p-type load transistor. It combines both a low mismatch senstivity in DC and a high gain in AC for the load. Common-mode feedback is applied to reduce the mismatch sensitivity of the input transistors. Furthermore, gatebackgate steering is used to increase the performance of all transistors. The amplifier has a gain of 15 dB. It is used in an AC-coupled 3-stage op amp. Subsequently, the design of a DC-connected 2-stage op amp is presented that applies the threshold voltage tuning technique to the input transistors in order to enable the DC connection of the stages. This 2-stage op amp has a gain of 20 dB.

The design of a 1st-order $\Delta\Sigma$ ADC is presented. It is built with an analog integrator, based on a 3-stage op amp. Furthermore, a high-gain comparator and a level shifter are used in the ADCs. The measured accuracy of the converter amounts to 26.5 dB in a 15.6 Hz signal bandwidth and at a clock speed of 500 Hz.

The implementation of a 1D and a 2D flexible capacitive touch sensor is discussed in this work. The 1D sensor has a triangular capacitive touch pad with a position dependent capacitance. In the sensor read-out, the charging current is copied and integrated on a fixed output capacitor. The sensor read-out has a sample rate of 1.5 kS/s. The simulated accuracy of the 1D sensor comes to 2 mm. The 2D flexible touch pad is based on the same working principle. An array of 4×4 pixels is measured in turns and the output signal shows the measured capacitance of the 16 pixels in a serial way. This output signal contains all the information about the position of the finger. According to simulations, the accuracy of the 2D sensor after interpolation is 0.3 mm.

The design of an organic Dickson DC-DC converter is presented. It is built with a ring oscillator, with buffer stages and with a Dickson converter core. The switches in the forward path have been implemented with diode-connected transistors. The presented converter reaches high and low output voltages of 48 V and −33 V respectively, at a power supply of 20 V. DC-DC converters are useful for biasing gates and backgates with high or low voltages using the threshold voltage tuning technique. The proof of concept of this technique is given by the DC-connected 2-stage op amp which is biased with the high bias voltage provided by the Dickson DC-DC converter. The power consumption of the converter is negligible compared to the power consumption of the building blocks which are or can be biased by this converter. As a result, the contribution of this building block is validated.

Acknowledgments

The authors would like to thank Polymer Vision for providing the organic electronics technology on foil in which all the presented circuits have been implemented and measured.

This research has been performed in the framework of the ESAT-imec research cooperation and in the HOLST centre imec-TNO research cooperation.

Contents

Abbreviations

1D, 2D, 3D	1-, 2-, 3-Dimensional
AC	Alternating Current
ADC	Analog-to-Digital Converter
BAN	Body Area Network
BG	Backgate
BGE	Bootstrapped Gain Enhancement
BW	Band Width
C-2C	Capacitor-Double-Capacitor Architecture
CMFB	Common-Mode Feedback
CMOS	Complementary Metal-Oxide-Semiconductor
CMRR	Common-Mode Rejection Ratio
CRT	Cathode-Ray Tube
DAC	Digital-to-Analog Converter
DC	Direct Current
DC-DC	DC to DC converter
EEF	Efficiency Enhancement Factor
GBG	Gate-Backgate Connection
GBW	Gain-Bandwidth Product
GHS	Globally Harmonised System of Classification and Labelling of Chemicals
GIZO	Gallium-Indium-Zinc-Oxide
IARC	International Agency for Research on Cancer
ITO	Indium-Tin Oxide
LG	Loop Gain
LUMO	Lowest Unoccupied Molecular Orbital
(MOS) FET	(MOS) Field Effect Transistor
MSB	Most Significant Bit
NAND	Not-AND Logic Gate
N/PMOS	n-type/p-type MOS transistor
(N/P)(O)TFT	n-type/p-type Organic Thin-Film Transistor
OLED	Organic Light-Emitting Diode

OSR	Oversampling Ratio
OVPD	Organic Vapor Phase Deposition
PC	Personal Computer
PEN	Polyethylene Naphthalate
PM	Phase Margin
PSD	Power Spectral Density
PVP	Poly-4-Vinyl Phenol
RFID	Radio Frequency IDentification
RGB	Red-Green-Blue
SAR	Successive Approximation Register
SN(D)R	Signal to Noise (and Distortion) Ratio
TVT	Threshold Voltage Tuning
VTE	Vacuum Thermal Evaporation
VTSR	Threshold Voltage Suppression Ratio
$\Delta\sum$	Delta-Sigma ADC
A	DC or pass-band gain
B	Up-scaling factor of a current mirror
C_A	Capacitance per area
C_G	Gate capacitance
C_L	Load capacitance
$C_{gd/gs(,k)}$	Gate-drain/gate-source capacitance (of transistor k)
C_k	Column k
$C_{ox(,t/b)}$	Oxide/insulator capacitance (of the top/bottom insulator)
C_{sens}	Sensor capacitance
I_L	Load current
I_{MOS}	Current in a MOS transistor
$I_{(S)D(,k)}$	Source-drain current of the transistor (transistor k)
$I_{SD,lin}$	Source-drain current in linear regime
$I_{SD,sat}$	Source-drain current in saturation regime
$I_{SD,sub}$	Source-drain current in subthreshold regime
$I_{d,0}$	Transistor current when VSG = VT
K'_p	Transistor constant in saturation
$L_{(k)}$	Transistor length (transistor k)
Met_i	Metal layer i
P_L	Power consumed in the load
P_{int}	Internal dissipated power
P_k	Power consumed in stage k
P_{tot}	Total power consumption
$R_{SD,lin}$	Source-drain resistance of a transistor in the linear regime
$R_{lin,i}$	Resistance of linear-transistor resistor i
R_{on}	On resistance
R_k	Row k
$R_{pent,i}$	Resistance of pentacene resistor i
T	Temperature
T_C	Critical temperature

V_{DD}	Supply voltage
V_E	Early voltage
V'_E	Early Voltage of the 4-pin OTFT
$V_{S,D,G,BG,B}$	Potential of the source, drain, gate, backgate or bulk node respectively.
V_{SS}	Ground voltage
V_T	Threshold voltage
V'_T	VT of the 4-pin OTFT
$V_{T,0}$	Initial threshold voltage
$V_{T,MOS}$	VT of the MOS transistor
$V_{XY(,k)}$	Potential difference between nodes X and Y (of transistor k)
$W_{(k)}$	Transistor width (transistor k)
a_k	Filter parameter of the kth filter
b	Number of bits of the internal ADC and DAC of the $\Delta\Sigma$ ADC
d	Distance between metal layers M1 and M2
f	Feedback factor
f_p	pole frequency
f_s	Sampling frequency
$g_{m(0)}$	(Initial) transiconductance of a transistor
$g_{m,k}$	Transconductance of transistor k
g_{sd}	Output conductance of a transistor, inverse of r_{sd}
k	Botzmann constant. $1.3806503 \cdot 10^{23}$ J/K
k_{out}	Ratio of output voltage and supply voltage
m_μ	Mean value of the mobility
n	Filter order of the $\Delta\Sigma$ ADC
n	Subthreshold slope factor
q	Elementary charge. $1.602176487 \cdot 10^{-19}$C
r_L	Load resistance
$r_{sd(,k)}$	Output resistance of a transistor (transistor k)
$r_{0(,k)}$	See $r_{sd(,k)}$
$t_{ox(,t/b)}$	Oxide/insulator thickness (of the top/bottom insulator)
γ	Technology dependent factor
ε_0	Electric permittivity of vacuum. 8.85 pF/m
$\varepsilon_{r(pvp)}$	Relative electric permittivity (of PVP)
η_P	Power efficiency
μ	Mobility
ξ	Sensitivity of V_T to V_{SBG}
π	Pi. 3.14159265358979323846264338327950288841...
σV_T	Variation of the V_T
$\sigma_{\mu(,adj)}$	Variation of the mobility (of adjacent transistors)
τ_{LR}	Time constant of an RL chain
ϕ_F	Fermi level
ϕ_k	Clock phase k

Chapter 1
Introduction

1.1 Prologue

In today's society one has his own mass-produced mobile phone that enables the user to connect with people all over the world from any location and at any moment. There is a large evolution of mobile devices and the present day's fashion-conscient consumer is inclined to follow these evolutions. As a result there is a very large and worldwide market for mobile communication devices and handheld computers, i.e. netbooks. The trend towards smaller and lighter devices conflicts with the interest for larger screens in these devices. In this context, it is worth it to have a look at organic electronics on foil. It is an electronics technology based on an organic semiconductor material instead of the more common semiconductors, e.g. silicon (Si). Circuits can in this technology be produced on top of a low-cost plastic foil.

The idea of a flexible and transparent plastic foil with integrated electronic circuits appeals to one's imagination. As opposed to the more typical Si chips, which are so small and have so many nanoscale components that their visibility has reduced over the years, the plastic chips are pretty large, in the order of a few cm^2. Even more, the idea of flexible displays fabricated on a plastic foil has encouraged several companies to develop prototypes of flexible display consumer devices such as foldable E-readers and smart phones. A few of these prototype consumer devices are shown in Fig. 1.1.

Flexible displays are an important yet not the only application of organic electronics. For instance, intelligent sensor systems which monitor critical parameters of the human body, are believed to be opportunities for this technology. This dissertation investigates the feasibility and the possibilities of such technology for the analog application domain.

H. Marien et al., *Analog Organic Electronics*, Analog Circuits and Signal Processing, DOI: 10.1007/978-1-4614-3421-4_1, © Springer Science+Business Media New York 2013

Fig. 1.1 **a** Prototype of the Readius E-reader by PolymerVision. **b** Prototype of a flexible display by Samsung

1.2 Origin and Driving Force

Only a short history of about 20 years precedes today's stage of development in organic electronics. In 1989, the first thin-film transistors (TFT) with organic semi-conductors were reported (Horowitz et al. 1989). Since that moment researchers all over the world have started to investigate and improve these organic devices. All sorts of organic materials have been experimented with. Pentacene has been adopted in the early years as a stable and reliable organic semiconductor for p-type transistors. In the course of time progress has been made and the first small organic integrated circuits were reported (Bonse et al. 1998; Vusser et al. 2006). Today, the technology is being developed for active-matrix backplanes of flexible displays (Gelinck et al. 2004; Huitema et al. 2008; Noda et al. 2010; Gelinck et al. 2010) and several companies have built and even presented prototypes of consumer devices with such flexible displays, e.g. E-readers and smart phones. Besides, the interest is growing to employ the organic electronics technology for several other applications, such as:

- Transponder chips for RFID applications (Myny et al. 2011; Heremans et al. 2010, 2011).
- Memory elements (Debucquoy et al. 2009).
- Various analog and mixed-signal circuits such as operational amplifiers, analog-to-digital (Marien et al. 2011), digital-to-analog converters (ADC and DAC) and DC–DC converters (Marien et al. 2011).
- Integrated thin-film sensors (Kawaguchi et al. 2005).

Mainly the analog circuits and to a certain extent also the sensors are dealt with in this dissertation.

1.3 Key Benefits and Challenges

The reason for the existence of organic electronics technology is twofold. They have a set of remarkable properties which create an alternative niche field of interest which is not reached with conventional silicon-based CMOS technology, and they are believed to reduce the future production cost of large-area circuits. The expected key benefits of organic electronics technology are:

- The low production temperatures.
- The possibility to use plastic substrates.
- The possibility to print circuits.

The first advantage is the processing temperature of this thin-film technology, which is typically kept below 150 °C, compared to production temperatures of CMOS chips that reach up to 1,000 °C. Such low processing temperatures enable the use of other materials than *Si* wafers as a substrate. Therefore, a low-cost material such as a plastic foil can serve as a substrate for organic electronic circuits. This plastic foil is a popular substrate since it has two main advantageous properties:

- It is low-cost, hence the surface area of circuits is less of a limit. More than that, large-area integrated applications, e.g. large-area sensor arrays, are affordable with this substrate.
- The plastic foil is a flexible material. As opposed to the CMOS chips with rigid *Si* substrates, circuits on plastic foil are foldable and even rollable. Indeed, a consumer device with a flexible screen will not break when it is dropped on the ground whereas a rigid glass screen of a traditional device certainly will.

There is another key benefit of organic electronics that stems from the low processing temperatures, i.e. the printability. The temperatures needed in the production of *Si* chips (1,000 °C) are so high that special heat-resistant and expensive materials are required. Owing to the reduced temperatures in organic electronics technology it is much easier to find materials which will not melt. As a result, the low temperatures enable the use of the standard printing techniques available in the well-known paper printers that are abundantly present in our home and work environments. In principle, it is possible to deposit all the required materials on the substrate by applying existing printing techniques and, in practice, the concept has been investigated and reported (Molesa et al. 2004). The printability is expected to reduce the production cost of plastic chips and the production delay would reduce to near-zero. At the moment, the printing of organic electronics circuits still suffers from a decreased performance. All circuits presented in this work have been produced with other deposition techniques than printing.

In combination with the key benefits of large-area and foldability, a roll-to-roll printing mechanism enables the large-area large-scale production of plastic circuits. With all these key benefits an infinite series of fancy consumer applications can be thought of, e.g. a printed movie on an A4 plastic foil, with integrated display, memory and user interface.

Despite all these benefits, and even though the application field of organic electronics technology looks promising, organic electronics technology also suffers from important drawbacks compared to *Si* based technology. In particular, the intrinsically low carrier mobility, the large variability of behavioral parameters and the low intrinsic gain of transistors (typically ~5) hamper circuit design. Moreover, the available component library is limited at present. As active elements only p-type transistors are available. Only very few early demonstrations of complementary logic have been reported. As passive components, only capacitors exist, while linear resistors are lacking. Resistive behavior is emulated by transistors biased in the linear regime.

Furthermore, the behavioral transistor parameters are affected by environmental influences at production time and at run time. As an example, degradation of the transistors occurs through bias stress, i.e. stress caused by the electric field in the transistor when the circuits are active. This effect is exacerbated by the presence of O_2 and H_2O in the air. Therefore, the challenge for organic circuits is that they have to be designed in such a way as to withstand a variation of the behavioral parameters, such as the threshold voltage V_T and the mobility μ.

1.4 Applications

The key benefits of organic electronics technology have led to several applications, which are in the making at the time this dissertation is written, and some more potential applications are being conceived of in a conceptual stage. They are all summarized in this section.

1.4.1 Flexible Active-Matrix Displays

Display devices have been evolving from the large cathode-ray tube (CRT) screens to the thin-film transistor (TFT) displays, i.e. the flat screens. This evolution towards integrated displays on a surface requires an active-matrix backplane. Every pixel in a display is driven by at least one transistor which determines the intensity of the light at that position. The collection of all these transistors, together with their interconnections, are called the active-matrix backplane of the display. Organic electronics technology is very well suited, it is actually designed, for building the active-matrix backplane of flexible displays. Two types of organic flexible displays have been reported in the literature: electrophoretic displays and OLED displays. The working principles of both of these organic displays are briefly discussed in this section.

1.4.1.1 Electrophoretic Displays

The working principle of flexible electrophoretic displays (Gelinck et al. 2004; Huitema et al. 2008) is presented in Fig. 1.2a. They are built with an electronic ink (E-ink) layer (Barrett et al. 1998). This is a layer with micro-capsules filled

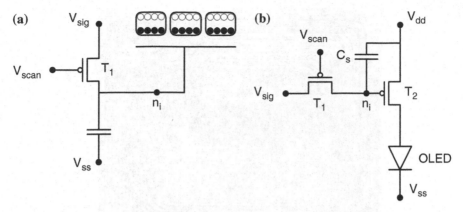

Fig. 1.2 **a** Simplified schematic view of the working principle of an electrophoretic display pixel. The voltage stored on the node n_i determines the color, either *black* or *white* (or an intermediate *gray-level*), of the pixel. **b** Simplified schematic view of the working principle of an OLED display. The voltage stored on the node n_i determines the current through T_2 and the illumination of the OLED. By putting *red*, *green* and *blue* OLED pixels next to each other full color images are created

with white and black microparticles. The white particles have a net negative charge whereas the black particles have a positive charge. By applying a negative or a positive voltage to node n_i, the pixel will turn either white or black. An active-matrix backplane with one transistor per pixel sets the value of each pixel. When the pixel is activated through V_{scan}, the node n_i will be set to the input signal V_{sig}. Since only one transistor is required, this backplane can be trouble-freely fabricated in a unipolar transistor technology. Electrophoretic displays have a gray-scale screen since the length of the activation pulse determines the level of black/white in the pixel (Gelinck et al. 2010). However, by using red-green-blue (RGB) color filters, the color palette can be extended towards full colors. An advantage of electrophoretic displays is that they are reflective displays, hence they can be read even in bright sunlight.

1.4.1.2 OLED Displays

Another approach to make organic flexible displays is to use organic light-emitting diode (OLED) displays driven by organic transistors, both of them integrated in the same technology. The active-matrix backplane is built with two transistors and a capacitor per pixel (Gelinck et al. 2010), as shown in Fig. 1.2b. When the pixel is activated through V_{scan}, the node n_i will be set to the input signal V_{sig} through T_1. Node n_i determines the current through T_2 and the OLED. Since both transistors are the same type (here p-type), this backplane can be implemented in a unipolar transistor technology. Each pixel drives one OLED that illuminates when a current flows. Contrarily to electrophoretic displays where ambient light is reflected on the display,

Fig. 1.3 Photograph of an RFID tag for theft prevention. The coil antenna is clearly visible

OLED displays generate light themselves. OLED displays offer the advantage of a wide viewing angle and ultra-high contrast.

1.4.2 RFID Tags

The concept of radio-frequency identification (RFID) tags is widespread in logistics and theft prevention. They are attached to an object and contain information about it. In fact, they are an alternative to bar codes that enables to put more information about the product on the tag. This information could specify the status, e.g. sold or not, or the model of the object, e.g. a serial number, and can be sent through an antenna to an RFID reader. Today's price of an RFID tag is between 0.05 and 0.2. A familiar example of an RFID tag for theft prevention is shown in Fig. 1.3. The largest part of the tag is a coil which functions as an antenna.

Since organic electronics technology is fabricated on a low-cost foil and has the property that it can be printed at large-scale, it meets the requirements of RFID tag applications. The area required for the foil is available at a very low cost and the full circuit can be integrated at once on the foil. This is expected to further reduce the cost per tag which is the driving force in every commercial application. Research on organic RFID tags on foil has been performed and digital RFID cores have been presented (Cantatore et al. 2007; Myny et al. 2011, 2012).

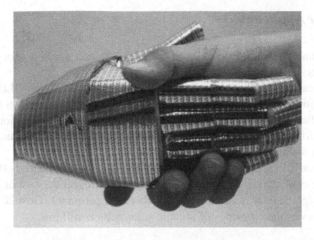

Fig. 1.4 Photograph of a robotic hand wrapped in an artificial skin with touch sensors (Kawaguchi et al. 2005)

1.4.3 Sensor Applications

As with flexible displays where an active matrix backplane determines the color and intensity of every pixel, distributed sensor systems driven by an active matrix are feasible. Expensive though such large sensor arrays were with classical technologies, they can now easily be fabricated at a low cost on a flexible plastic foil. A good example of an application of these sensor arrays is given by a distributed pressure sensor (Kawaguchi et al. 2005). An active matrix of pixels is enhanced by a rubbery material with a pressure-dependent resistivity. This sensor is implemented on foil hence it acts as some kind of artificial skin. It can be employed to add a skin to the moving parts of a robotic machine, to evoke pressure-dependent signals in these parts when they are touching an object. The example of a robotic hand is shown in Fig. 1.4. In this way organic electronics can create an added value for the whole robot industry.

Besides this idea, i.e. adding another layer with special properties, it is also possible to employ the behavior of the organic transistors themselves so as to make fully integrated sensor systems. The sensitivity of organic thin-film transistors to all kinds of external influences has been reported, e.g. temperature, pressure and chemicals. With a creative usage of these sensitivities, sensor circuits can be fully integrated in a technology. An interesting example is found in chemical sensors that measure the acidity of milk or whine (Subramanian et al. 2006). The application of such sensors in smart sensor systems is elaborately discussed in Sect. 1.5.

1.4.4 Other Applications

Other opportunities for organic electronics can be found, which do not fit into the previous Sects. 1.4.1–1.4.3.

One such example is flexible lighting. For this application large OLEDs are deposited on a foil and used as a light source. They can be easily cut in the desired shapes. The flexibility of the foil makes it useful in safety applications, e.g. for marking the difficult steps of a stairway. Another more fancy application could be light-emitting wallpaper used to illuminate a room.

Another example application is found in printed organic solar cells (Chuo et al. 2011; Shafiee et al. 2008). By printing them on a plastic foil the cost of those solar cells can be significantly reduced from €10 to €0.03 per panel (Oost 2011). At the moment, however, the efficiency of organic solar cells is still low.

1.5 Organic Smart Sensor Systems

The subject of this book is the exploration of organic smart sensor systems. This section introduces the concept of smart sensor systems and subsequently proposes an architecture for its implementation in organic electronics technology.

1.5.1 Smart Sensor Systems

Smart sensor systems are circuits that combine a certain digital intelligence, e.g. an RFID tag, with the ability to monitor certain parameters through sensors, and, possibly, to drive an actuator, of which the behavior is determined by the status of the system. They are applicable in several kinds of monitoring, e.g. continuous health or food monitoring, where they can save the high cost of a professional person. In a medical environment this can mean the monitoring of physiological parameters of sportsmen and elderly people. Similarly a smart sensor system could monitor a piece of meat during the time interval between production and consumption, in order to check if the temperature of conservation has been low enough at all times.

Smart sensor systems make use of both analog and digital circuits. The digital core is enhanced with sensors which receive an analog signal from the environment. As a result, an analog sensor read-out and an analog-to-digital conversion block are required as an interface between both. The analog building blocks as well as the sensors are the subject of this research.

1.5.2 Proposed Architecture

Figure 1.5 shows the proposed architecture of an organic smart sensor system. The system consists of five building blocks, four of which are analog. Their functions

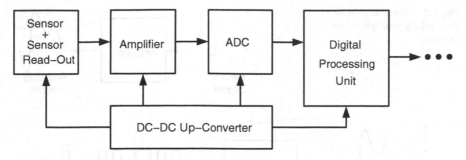

Fig. 1.5 Block diagram of an organic smart sensor system

Fig. 1.6 Representation of
the working principle of a
sensor

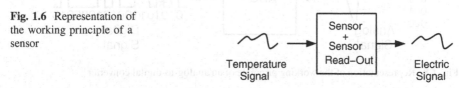

are briefly discussed in Sects. 1.5.2.1–1.5.2.4. Chapters 3–6 of this dissertation are devoted to one of these analog building blocks each.

1.5.2.1 The Sensor

The representation of the working principle of a sensor is shown in Fig. 1.6. The basic functionality of a sensor is to convert an external analog signal into an electric signal which is either a current or a voltage. This general description can be applied to every sensor, e.g. a temperature sensor converts the temperature signal as a function of time into an electric time-signal, or likewise a chemical sensor converts a chemical concentration into a current or voltage signal. Two important characteristics of sensors should be mentioned.

Firstly, sensors are preferably very linear, i.e. the electric signal is proportional to the environmental signal. This facilitates the interpolation of the sensor afterwards.

Secondly, a sensor always interferes with the environment in a certain way, e.g. when temperature is measured a little amount of heat is extracted, which modifies the original temperature. A sensor should always be designed in such a way that this interference with the external signal is negligible. Therefore, sensors often give only a very small electric signal at their output. This forms the direct reason for the existence of the next building block, i.e. the amplifier.

1.5.2.2 The Amplifier

Since the output signal of the sensor is typically of low amplitude, it must be amplified through a front-end amplifier. The output of this amplifier is the same signal, yet

Fig. 1.7 Representation of
the working principle of an
amplifier

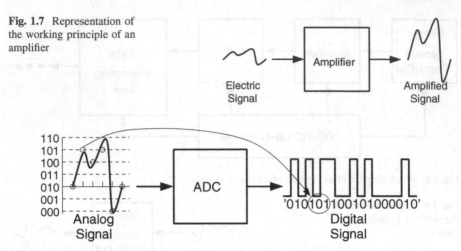

Fig. 1.8 Representation of the working principle of an analog-to-digital converter

amplified into a large swing signal in a linear way, as can be seen in Fig. 1.7. Although
the amplifier is the most basic analog building block, its implementation in organic
electronics technology is not obvious.

1.5.2.3 The Analog-to-Digital Converter

The function of the analog-to-digital converter ADC is explained in Fig. 1.8. An
analog input signal, continuous in amplitude and continuous in time, is converted
to a digital signal, represented by discrete values, i.e. logic 1's and 0's, in discrete
time steps. In the ADC a set of quantization levels for the analog input signal can be
employed. Every quantization level corresponds to a digital code, in this example a
3-bit word. At fixed time steps, the analog signal is round off to the closest quantized
level and the corresponding digital word is stored. The digital output of the ADC
represents the consecutive digital words in a serial way.

1.5.2.4 The Digital Core

The digital core of the smart sensor system is the place where the sensed information
is processed and stored. In this building block the added value of the smart sensor
system, i.e. a certain intelligence, is created. The functionality depends on the application. One implementation could be that the system actuates a red led that tells
the consumer that the temperature of a piece of meat has gone too high and that is
should no longer be consumed. Another implementation could be that such a system
stores monitored information about the heart beat of a patient and sends it wirelessly

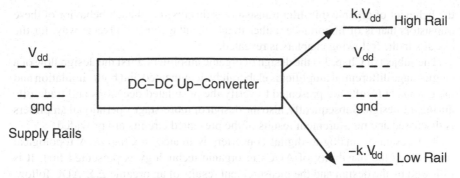

Fig. 1.9 Representation of the working principle of a DC–DC converter. V_{dd} and *gnd* represent the supply and the ground voltage respectively

to the base station of a body area network (BAN). The implementation of this digital building block is beyond the scope of this research and it is not further discussed.

1.5.2.5 The DC–DC Converter

The working principle of a DC–DC up-converter is visualized in Fig. 1.9. This circuit generates DC voltages that are either above the supply voltage rail or below the ground voltage rail. Each generated output voltage can be used as a power supply rail or only as a bias voltage.

The DC–DC converter is an analog building block that is typically not employed in a sensor read-out path. In the organic electronics technology the high and low output voltages are very useful as bias voltages. These bias voltages can improve the quality of other building blocks, both analog and digital. The DC–DC converter has a supportive function towards the other building blocks in this work. A thorough discussion of this topic is given in Chap. 6.

1.6 Outline of This Work

The objective of the research presented in this work was to investigate the feasibility of analog circuit design in the available organic electronics technology. Such a broad objective has created an enormously large research field of interest, always with overtones of analog circuit design. It has led to the exploration and the development of several distinct typical analog circuits which are presented. They are united in an umbrella application in the proposed block diagram of the organic smart sensor system, depicted in Fig. 1.5. This idea forms the backbone of this book.

Chapter 2 introduces the reader into the world of organic electronics. Briefly, the available technologies and the production techniques are explored. Furthermore, the behavior of organic thin-film transistors is disclosed and each behavior of these transistors that is of importance, either in an advantageous or adverse way, for the circuits in the following chapters is revealed.

The subject of Chap. 3 is the design of organic amplifiers. First the design flow of a single-stage differential amplifier is elaborately documented and both simulation and measurement results are presented to verify the postulated decisions and trade-offs during the design. Subsequently, also the design of more-stage operational amplifiers is discussed and measurement results of the presented circuits are provided.

The design of analog-to-digital converters is treated in Chap. 4. A topological study about the feasibility of ADCs in organic technology is presented first. It is followed by the design and the measurement results of an organic $\Delta\Sigma$ ADC follow.

Subsequently, the integration of sensors in the existing technology is investigated in Chap. 5. In this chapter the implementation of $1D$ and $2D$ capacitive touch sensors is treated and measurement results are presented.

Next, Chap. 6 explores the domain of organic DC–DC converters. An extensive theoretical comparison between different topologies is given first, followed by the presentation of a few chip designs of a Dickson DC–DC converter.

Finally, the general conclusions of this dissertation are drawn in Chap. 7. Also, the main contributions of this work are summarized and the impetus is given to future work.

References

Barrett C, Albert JD, Yoshizawa H, Jacobson J (1998) An electrophoretic ink for all-printed reflective electronic displays. Nature 394(6690):253–255

Bonse M, Thomasson DB, Klauk H, Gundlach DJ, Jackson TN (1998) Integrated a-Si:H/pentacene inorganic/organic complementary circuits. In: International electron devices meeting 1998 IEDM '98 technical digest, pp 249–252.

Cantatore E, Geuns TCT, Gelinck GH, van Veenendaal E, Gruijthuijsen AFA, Schrijnemakers L, Drews S, de Leeuw DM (2007) A 13.56-MHz RFID system based on organic transponders. IEEE J Solid-State Circuits 42(1):84–92.

Chuo Y, Omrane B, Landrock C, Aristizabal J, Hohertz D, Niraula B, GrayliS V Kaminska B (2011) Powering the future: integrated, thin, flexible organic solar cells with polymer energy storage. IEEE Des Test Comput 28(99):1

De Vusser S, Steudel S, Myny K, Genoe J, Heremans P (2006) A 2V organic complementary inverter. In: IEEE international solid-state circuits conference ISSCC 2006, Digest of technical papers, pp 1082–1091.

Debucquoy M, Rockele M, Genoe J, Gelinck GH, Heremans P (2009) Charge trapping in organic transistor memories: on the role of electrons and holes. Org Electron 10(7):1252–1258

Gelinck G, Heremans P, Nomoto K, Anthopoulos TD (2010) Organic transistors in optical displays and microelectronic applications. Advanced materials (Deerfield Beach, Fla.) 22(34):3778–3798.

Gelinck GH, Edzer H, Huitema A, van Veenendaal E, Cantatore E, Schrijnemakers L, van der Putten JBPH, Geuns TCT, Beenhakkers M, Giesbers JB, Huisman B-H, Meijer EJ, Benito EM, Touwslager FJ, Marsman AW, van Rens BJE, de Leeuw DM (2004) Flexible active-matrix displays and shift registers based on solution-processed organic transistors. Nat Mater 3(2):106–110

Heremans P, Dehaene W, Steyaert M, Myny K, Marien H, Genoe J, Gelinck G, van Veenendaal E (2001) Circuit design in organic semiconductor technologies, In: Proceedings of the ESSCIRC (ESSCIRC) 2011, pp 5–12.

Heremans P, Myny K, Marien H, van Veenendaal E, Steudel S, Steyaert M, Gelinck G (2010) Application of organic thin-Film transistors for circuits on flexible foils. In: Proceedings of international display workshop, Fukuoka, 2011.

Horowitz G, Fichou D, Peng X, Xu Z, Garnier F (1989) A field-effect transistor based on conjugated alpha-sexithienyl. Solid State Commun 72(4):381–384

Huitema HEA et al. (2008) Rollable displays: the start of a new mobile device generation. In: 7th annual USDC flexible electronics and display conference, 2008.

Myny K, Rockel M, Chasin A, Pham D-V, Steiger J, Botnaras S, Weber D, Herold B, Ficker J, van der Putten B, Gelinck GH, Genoe J, Dehaene W, Heremans P (2012) Bidirectional communication in an hf hybrid organic/solution-processed metal-oxide rfid tag, In: IEEE international solid-state circuits conference digest of technical papers (ISSCC) 2012, Accepted for 2012.

Kawaguchi H, Someya T, Sekitani T, Sakurai T (2005) Cut-and-paste customization of organic FET integrated circuit and its application to electronic artificial skin. IEEE J Solid-State Circuits 40(1):177–185

Marien H, Steyaert M, van Veenendaal E, Heremans P (2011) DC-DC converter assisted two-stage amplifier in organic thin-film transistor technology on foil. In: Proceedings of the ESSCIRC 2011, pp 411–414.

Marien H, Steyaert MSJ, van Veenendaal E, Heremans P (2011) A fully integrated $\Delta\Sigma$ ADC in organic thin-film transistor technology on flexible plastic foil. IEEE J Solid-State Circuits 46(1):276–284

Molesa SE, Volkman SK, Redinger DR, Vornbrock AdF, Subramanian V (2004) A high-performance all-inkjetted organic transistor technology. In: IEEE international electron devices meeting 2004, IEDM technical digest, pp 1072–1074.

Myny K, Beenhakkers MJ, van Aerle NAJM, Gelinck GH, Genoe J, Dehaene W, Heremans P (2011) Unipolar organic transistor circuits made robust by Dual-Gate technology. IEEE J Solid-State Circuits 46(5):1223–1230

Noda M, Kobayashi N, Katsuhara M, Yumoto A (2010) Ushikura S (2010) A rollable AM-OLED display driven by OTFTs. Proc SID 41:710–713

Shafiee A, Salleh MM, Yahaya M (2008) Fabrication of organic solar cells based on a blend of donor-acceptor molecules by inkjet printing technique. In: IEEE international conference on semiconductor electronics ICSE 2008, pp 319–322.

Subramanian V, Lee JB, Liu VH, Molesa S (2006) Printed electronic nose vapor sensors for consumer product monitoring. In: Solid-State Circuits Conference, 2006. ISSCC 2006. Digest of Technical Papers. IEEE, International, pp 1052–1059.

Van Oost J (2011) Universiteit hasselt vindt printbaar zonnepaneel uit http://www.zdnet.be. 2011

Chapter 2
Organic Thin-Film Transistor Technology: Properties and Functionality

The field of organic electronics is an example of a technology which explores a totally different path in integrated electronics with a new and specific niche application field including flexible displays, organic lighting, and smart systems. This application field is the driving force for the development of the technology and the presentation of the first thin-film transistor with organic semiconductors in 1989 (Horowitz et al. 1989) has nowadays transformed into a whole printed electronics industry with a $2.2 Billion business in 2011 and a predicted $44 Billion business for 2021 (IDTechEx 2011).

Organic electronics technology is still in its infancy and suffers from a series of technological and behavioral parameter variations caused by the production process and by the environment, e.g., temperature and humidity. However, organic technology also displays interesting features. Of course, the flexibility and the price of the plastic substrate as well as the experimentally proven printability (Molesa et al. 2004) are the technological enablers of the application field. Furthermore physical features are present at the transistor level that enable circuit techniques which are not suited for Si standard CMOS. The so-called backgate of the organic transistor enables a series of techniques that improve circuit behavior. The design of a circuit in organic electronics technology is different from circuit design in Si standard CMOS and needs to be based on the specific technological background with related working principles, advantages, and challenges. Therefore in this chapter a closer look on organic electronics technology is given.

First, an introduction is given about thin-film transistors in Sect. 2.1. Consequently, Sect. 2.2 uncovers the chemical, physical, and electrical properties of the organic semiconducting materials. Next, in Sect. 2.3 an overview of the applied deposition techniques is presented. The technology in which most of the circuits in this dissertation are fabricated is explained in Sect. 2.4 where special attention is given to the standard 3-contact and the extended 4-contact transistor architectures. Furthermore, the advantages of the 4-contact transistor architecture are explained through practical examples. Section 2.5 focuses on the non-ideal behavior of organic transistors and explores the dominant degradation effects which are present in organic

H. Marien et al., *Analog Organic Electronics*, Analog Circuits and Signal Processing,
DOI: 10.1007/978-1-4614-3421-4_2, © Springer Science+Business Media New York 2013

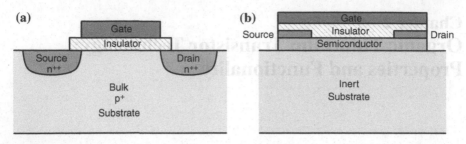

Fig. 2.1 **a** The cross-section of an NMOSFET which is embedded in a Si substrate. **b** The cross-section of a TFT which is deposited on top of an inert substrate. Both the MOSFET and the TFT are field-effect transistors

electronics technology. The presence of transistor models which are physically rather than behaviorally inspired is discussed in Sect. 2.6. Consequently the implementation of the passive components, i.e., resistors, capacitors, and inductors, in the organic electronics technology is treated in Sect. 2.7. Finally, Sect. 2.8 summarizes this chapter.

2.1 Thin-Film Transistors

As indicated by the title, the transistors in the applied technology are thin-film transistors. The latter differs from the more common MOSFET transistors in Si standard CMOS both in a physical and a behavioral way. The cross-section of an n-type MOSFET transistor (NMOS) and a thin-film transistor (TFT) is presented in Fig. 2.1. The NMOS transistor is embedded on a p^+-doped Si substrate. The two n^{++}-doped areas represent the source and the drain of the NMOS and the transistor channel is situated between them. A gate contact, typically manufactured in polysilicon, is deposited on top of the channel. When the gate is biased positively a space charge region is formed and the channel conducts electrons (e^-) from the source to the drain while the transistor channel is in the inversion mode, i.e., the channel conducts electrons e^-. MOSFET transistors also have a bulk contact which is the area close to the transistor channel. The bulk of this NMOS is connected to the substrate, so the bulk contacts of all NMOS transistors are connected. This can, if necessary, be overcome through a triple well but that is beyond the scope of this dissertation. p-type MOSFET transistors (PMOS) are fully complementary with NMOS transistors. Their working principle is based on the conduction of holes.

In a thin-film transistor (TFT) the semiconductor is sufficiently thin so that it can be fully depleted by the gate. It has ohmic drain and source contact regions. Furthermore, it has no substrate contact. The basic architecture of a TFT is similar to a MOSFET transistor, since the gate contact is electrically separated from the channel by an insulator. Both are field-effect transistors (FET).

The current of a p-type TFT in saturation is given by Eq. (2.1):

$$I_{SD,sat} = \mu C_{ox} \times \frac{W}{L}(V_{SG} - V_T)^2 \times \left(1 + \frac{V_{SD}}{V_E \times L}\right), \qquad (2.1)$$

where μ is the carrier mobility, C_{ox} the physical insulator capacitance, V_T the threshold voltage, and V_E the Early voltage of the transistor. This equation is identical to the equation of the saturation current in a PMOS. Moreover, the subthreshold current of a TFT can be well described by Eq. (2.2), showing an exponential dependency on V_{SG}, similar to the subthreshold current in MOSFETs, although the physical origin of the subthreshold region is different for both types of transistors. In a MOSFET transistor the subthreshold region is related to a diffusion current, whereas in a TFT it is related to the pinch-off of both the drift current and the hopping mechanism through the trap states of the density of states (DOS) (Oberhoff et al. 2007b).

$$I_{SD,sub} = I_{d,0} \times \frac{W}{L} \times e^{\frac{V_{SG}}{nkT/q}} \qquad (2.2)$$

In this equation, $I_{d,0}$ is the transistor current when $V_{SG} = V_T$ and n a biasing voltage-dependent parameter that determines the steepness of the subthreshold slope.

TFTs have the advantage that they are not embedded in the same substrate and hence there are no leakage paths in the insulating substrate that propagate noise. Furthermore, it is possible to make a fourth contact in a TFT that is more or less the counterpart of the bulk contact in a MOSFET. This is further discussed in Sect. 2.4.2.

The physical architecture of a TFT requires at least four layers: one for the semiconductor, one metal layer for the source and drain contacts, one for the insulator, and finally a metal layer for the gate. With these layers there are four different ways to build a TFT, which are visualized in Fig. 2.2. Which one of these architectures is used is dictated by the specific production process. In the next section the state-of-the-art organic semiconductors and their characteristics are described.

2.2 Organic Semiconductors

The first organic thin-film transistor has been reported in 1989 (Klauk 2006; Horowitz et al. 1989). Since then the research interest on organic transistors has more and more developed and along the way pentacene has come out as a stable hence often used organic semiconductor that has been heavily investigated and still is.

In this section the most developed and most used organic semiconductor, pentacene, is presented and the working principle is elucidated. Afterwards an overview is given of other promising organic semiconductors.

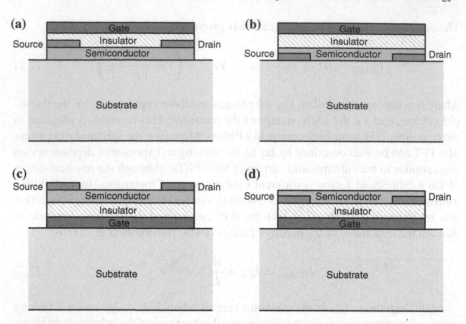

Fig. 2.2 The four implementations of a four-layer TFT. **a** The *top-gate* architecture with source and drain contacts on *top* of the semiconductor. **b** The *top-gate* architecture with source and drain contacts *below* the semiconductor. **c** The *bottom-gate* architecture with source and drain contacts on *top* of the semiconductor. **d** The *bottom-gate* architecture with source and drain contacts *below* the semiconductor

2.2.1 Pentacene

Pentacene ($C_{22}H_{14}$) is an organic small molecule, more specifically a polycyclic aromatic hydrocarbon that consists of five fused benzene rings, as can be seen on the drawing in Fig. 2.3a. The benzene rings are fused in a linear chain and form a planar molecular structure in which the delocalized electrons can move freely. An atomic force microscopic photograph of the pentacene molecule is shown in Fig. 2.3b. The uncomplicated structure of the molecule enables a tight molecule stacking in a planar herringbone crystal structure, as visualized in Fig. 2.3c. This two-dimensional herringbone crystal structure is further stacked in the third dimension.

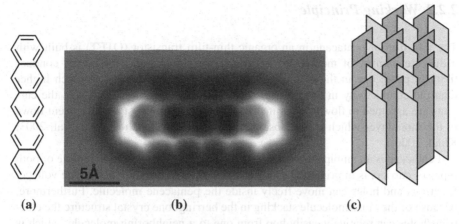

(a)   (b)   (c)

Fig. 2.3 **a** The chemical structure of pentacene consists of five fused benzene rings. **b** Atomic force microscopic view of a single pentacene molecule (Gross et al. 2009). **c** The pentacene molecules stack in a two-dimensional herringbone crystal structure

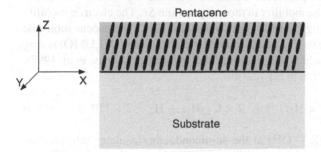

Fig. 2.4 Cross-sectional view of a pentacene thin-film layer. Pentacene tends to grow in monolayers which are located in the X–Y plane and growing in the Z direction on top of a substrate. This is of course an ideal representation without defects, dislocations, or grain boundaries

Toxicological Properties of Pentacene
The documentation of the toxic properties of pentacene is not easily found in the literature. The most accurate information on pentacene is given in the data sheets of commercially available pentacene (Sigma-Aldrich 2010). According to the Globally Harmonised System of Classification and Labelling of Chemicals (GHS) the product is not a dangerous substance. According to the International Agency for Research on Cancer (IARC) no component of this product (pentacene) present at levels greater than or equal to 0.1 % is identified as probable, possible, or confirmed human carcinogen

2.2.2 Working Principle

The thin film of pentacene in an organic thin-film transistor (OTFT) is built with a discrete number of monolayers, as can be seen in Fig. 2.4. Since the conductivity in pentacene in the X–Y plane, i.e., inside one monolayer, is much higher than the conductivity in the Z direction, i.e., in between the monolayers the currents are assumed to flow in a laminar way. Furthermore, most of the current flows in the monolayer which is the closest to the insulator (Sancho-Garca et al. 2003; Klauk 2006).

Pentacene is an ambipolar semiconductor since the accumulation regime of both types of carriers is in principle possible, depending on the sign of the gate voltage. Electrons and holes can move freely inside the pentacene molecule. Furthermore, because of the close molecule stacking in the herringbone crystal structure the electrons/holes can relatively easily hop from one to a neighboring molecule, which is actually a quantum mechanical tunneling effect (Gelinck et al. 2005). Grain boundaries are the most critical regions for carrier hopping and these pull the mobility down. The hole mobility in a pentacene transistor is around $0.1–1\,cm^2/Vs$ which is about three decades below the mobility in monocrystalline Si. The electron mobility is much lower and even negligible in air for two reasons. First, a pentacene molecule that possesses an electron in its lowest unoccupied molecular orbital (LUMO) is very reactive to H_2O molecules which are present in the ambient air (Leeuw et al. 1997), according to the redox reaction in Eq. (2.3):

$$2 \times C_{22}H_{13}^- + 2 \times H_2O \;\rightleftharpoons\; 2 \times C_{22}H_{14} + H_2 + 2 \times OH^- \qquad (2.3)$$

Second, hydroxyl groups $(R - OH)$ at the semiconductor–insulator interface are believed to act like e^- traps which immobilize the negative charge carriers (Chua et al. 2005). Pentacene is therefore considered, as most other organic semiconductors, to be a p-type semiconductor only. When the pentacene transistor is biased on, holes are accumulated in the channel near the semiconductor–insulator interface which starts to conduct current. In practice, pentacene is considered as a unipolar semiconductor.

2.2.3 Alternative Organic Semiconductors

The research for organic semiconductors is performed by chemists and material engineers all over the world. Without aiming to give a full material and chemical analysis, a few of the most promising engineered organic semiconductors are summarized in this section.

2.2.3.1 Buckminsterfullerene

Buckminsterfullerene, better known as Bucky balls or C_{60}, is a spherical molecule built with carbon atoms only. This molecule, which consists of 20 hexagons and 12

(a) **(b)**

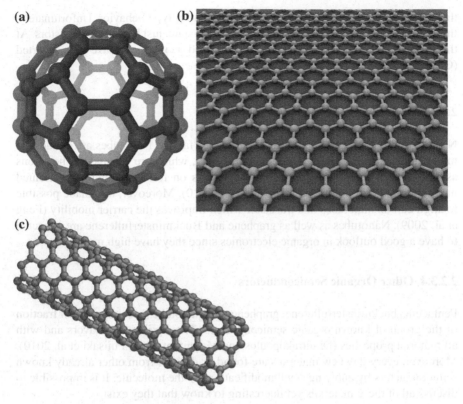

(c)

Fig. 2.5 Graphical representation (Wikipedia 2011) of the chemical structure of (**a**) Buckminster-fullerene, (**b**) a graphene sheet, and (**c**) a carbon nanotube

pentagons, is visualized in Fig. 2.5a. The electrical properties are, similar to pentacene, caused by delocalized electrons in the molecules, yet the C_{60} transistors are known to exhibit n-type transistor behavior. C_{60} transistors perform with a mobility that is in the same order of magnitude as pentacene transistors and therefore both can be combined in an organic complementary technology (Na et al. 2008; Bode et al. 2010).

2.2.3.2 Graphene

Graphene, shown in Fig. 2.5b, is a fascinating material that consists of carbon atoms arranged in hexagons in a 2D plane. The production of single graphene layers has been reported only in 2004 (Novoselov et al. 2004). The interest for this material has been the driving force for researchers and nowadays transistors with this organic semiconductor have been reported (Schwierz 2010) and very high mobilities have been measured (up to 15,000 cm²/Vs). Graphene transistors have the advantage that they are ambipolar, i.e., they show both n-type and p-type behavior. Unfortunately, this also means that they cannot be biased in an off-state, just like other transistors. At the moment, only a few small digital circuits based on graphene have been reported (Chen et al. 2011).

2.2.3.3 Nanotubes

Nanotubes are just like C_{60} part of the fullerenes family. Nanotubes are cylindrical molecules that also consist of carbon atoms only, which are arranged in hexagons as can be seen in Fig. 2.5c. Thin-film transistors on foil with randomly oriented nanotubes have been reported (Jang and Ahn 2010). Moreover, it is made possible to align carbon nanotubes in a transistor which improves the carrier mobility (Feng et al. 2009). Nanotubes as well as graphene and Buckminsterfullerene are expected to have a good outlook in organic electronics since they have high mobilities.

2.2.3.4 Other Organic Semiconductors

Pentacene, buckminsterfullerene, graphene, and nanotubes are only a small fraction of the group of known organic semiconductors. They exist in all flavors and with all material properties (Dimitrakopoulos and Malenfant 2002; Fujisaki et al. 2010). Moreover, every day new materials are found or derived from other already known semiconductors by applying small modifications to the molecule. It is impossible to discuss all of these materials yet interesting to know that they exist.

2.2.3.5 Inorganic Oxide Semiconductors

Another group of materials that show semiconductor behavior are the inorganic oxides. Although these materials are not classified in the family of organic materials, they perform very well in thin-film transistors. Gallium-Indium-Zinc-Oxide (GIZO) is such an alloy that is at the moment often used to make n-type TFTs. Their carrier mobility is typically 10 times higher than in pentacene transistors. They are mentioned here because they are sometimes combined with p-type organic semiconductors, like pentacene, to form a complementary technology on foil (Na et al. 2008).

2.3 Production Techniques

One of the most important arguments in favor of organic electronics is the printability of the circuits. The low processing temperatures enable the use of flexible plastic substrates which in turn, at least in principle, enables roll-to-roll printing of the circuits. During the research phase the organic electronics technologies are produced with other techniques, such as vacuum thermal evaporation and vapor-phase deposition.

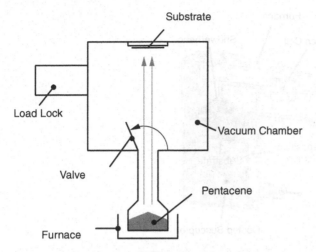

Fig. 2.6 Schematic view of the VTE deposition technique for pentacene thin films

The production process of the pentacene technologies (3 metal layers) has a low complexity compared to the production process of *Si* standard CMOS (5–10 metal layers). The critical layer in this process is the pentacene layer. The level of ordering in the crystals, the grain size of the pentacene grains, and the presence of lattice defects all have an influence on the transistor parameters, such as the mobility (μ) and the threshold voltage (V_T). In this section, the different available deposition techniques for the deposition of pentacene and their advantages are briefly considered.

2.3.1 Vacuum Thermal Evaporation

One technique which is used for the growth of a thin-film pentacene layer is vacuum thermal evaporation (VTE). For this technique the substrate is brought into a vacuum chamber. A source cell with pentacene powder is heated so thermal evaporation takes place. Because of the vacuum in the chamber the sublimated pentacene molecules have a large travel distance. As a result, they keep to a straight path. Some of the molecules travel straight toward the substrate. The other molecules which do not travel straight to the substrate collide with and stick to the walls of the vacuum chamber. A valve is present to start and terminate the growth process of the pentacene thin film. The morphology of the thin film is determined by the vacuum pressure, the heating temperature of the pentacene source cell, and the temperature of the substrate (Verlaak et al. 2003; Wang et al. 2011).

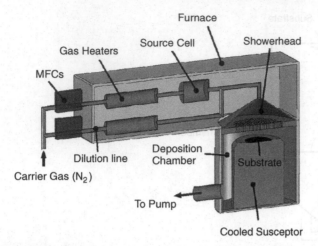

Fig. 2.7 Working principle of the organic vapor-phase deposition (OVPD) technique for pentacene thin films (Rolin et al. 2006)

2.3.2 Vapor-Phase Deposition

Another technique for the deposition of a thin-film layer of organic molecules, such as pentacene, is organic vapor-phase deposition. The basic principle of this technique is presented in Fig. 2.7 and is suitable for high volume production (Rolin et al. 2006). The pentacene powder is stored in a source cell. An inert N_2 gas flow is generated which is heated up to 270 °C and directed over the pentacene. The pentacene molecules come into the N_2 through sublimation and the gas goes through a showerhead which is optimized for a uniform pentacene flow in the deposition chamber. In the deposition chamber a susceptor is present with the substrate placed on top. The susceptor is cooled down to 20–50 °C and the pentacene molecules in the gas attach to the substrate, now through condensation. The structure and the morphology of the pentacene layer are highly influenced by the process parameters, such as temperature and the gas flow. By optimizing these parameters a uniform deposition rate and a good morphology in the pentacene layer are pursued (Rolin et al. 2006).

2.3.3 Solution-Based Deposition

When a material is deposited from a solution rather than from the gas state, this is called solution-based deposition. This technique is advantageous over vacuum techniques such as VTE since it does not require a vacuum chamber. Furthermore solution-based deposition is an enabler for printed circuits. For the technique of solution-based deposition a pentacene precursor is dissolved in a solution (Herwig and Mllen 1999). The deposition is carried out by coating the substrate with the

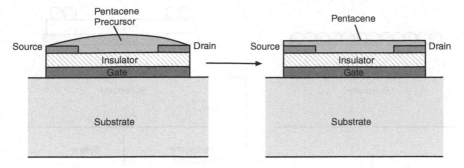

Fig. 2.8 Working principles of the solution-based deposition technique for pentacene thin films

molecule. Subsequently the chemical reaction that converts the pentacene precursor into pentacene takes place. It is triggered by an external parameter, typically a temperature step. The principle of solution-based printing is visualized in Fig. 2.8.

2.3.4 Inkjet Printing

A lot of effort is being spent by research groups all over the world to the printing of organic circuits. This technique is the enabler of large-scale and large area applications. Several printing techniques exist, yet inkjet printing is believed to be the best applicable technique for the printing of organic circuits.

An example of an inkjet printing technique is described in Molesa et al. (2004). Figure 2.9 visualizes the consecutive steps of this inkjet printing process. First the gate electrode is deposited from Au ink based on nanoparticles or an Au precursor. An annealing step converts this layer to a continuous Au film. Subsequently an insulator, here poly-4-vinyl phenol (PVP), is coated on top of the gate. The source and drain contacts are processed in the same way as the gate layer, i.e., by deposition of gold particles or a precursor followed by an annealing step. Finally the pentacene precursor is printed and converted to pentacene.

2.3.5 Self-Aligned Transistors

All previously mentioned deposition techniques suffer from misalignment issues due to the flexible nature of the substrate. Therefore, an overlap of several micrometers is employed which ensures the transistor performance. However, through this safety margin in terms of overlap a lot of overlap capacitance is created between the gate and the drain and between the gate and the source in the transistor architecture. As can be seen in Fig. 2.2 there is a 100 % overlap between the source and the gate contacts, as well as between the drain and the gate contact which slows down the

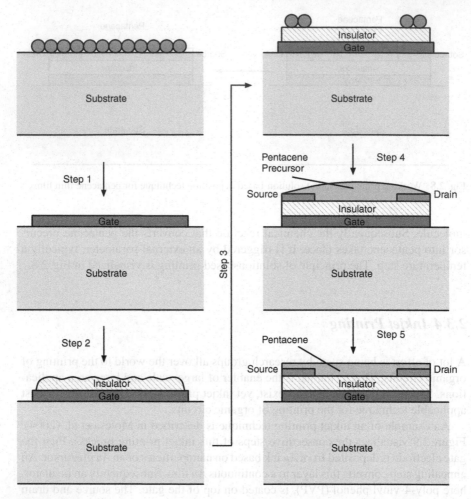

Fig. 2.9 Step-by-step inkjet printed deposition of pentacene thin films (Molesa et al. 2004)

circuit speed. By applying techniques that overcome the misalignment issue, self-aligned transistors can be fabricated without a gate-source or gate-drain overlap. This is typically done by using one of the two materials which are to be aligned as the mask for the deposition of the other material. Figure 2.10 gives an example that explains the production of a self-aligned top-gate thin-film transistor (Morosawa et al. 2011).

Fig. 2.10 Step-by-step process for a self-aligned thin-film transistor (Morosawa et al. (2011)). A reactive *Al* layer is deposited that reacts with the semi-conductor layer. Self-aligned source and drain contacts are formed without overlap with the gate. The remaining *Al* reacts with O_2 and forms an Al_2O_3 passivation layer

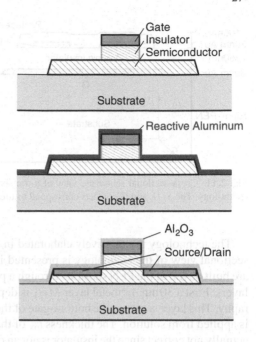

2.4 Technology in This Work

The existing organic electronics technologies are all very novel and undergo processing changes along the road that improve the transistor behavior. Consequently, it is often difficult to compare circuits fabricated in different technologies. Therefore, it is important to start this dissertation about integrated organic circuits with a brief overview on the applied technology and its features. Section 2.4.1 presents the standard 3-contact transistor technology. In Sect. 2.4.2 the modified 4-contact transistor is heavily considered, and finally in Sect. 2.4.3 a few of the features that are enabled by the 4-contact transistor are highlighted.

2.4.1 3-Contact OTFT

The technology applied for the designs in this work is a dual-gate pentacene thin-film transistor implemented on a plastic substrate. The substrate is a 25 μm plastic foil with a 150 mm diameter fabricated in polyethylene naphtalate (PEN), a low-cost polymer which is used in a whole set of applications, reaching from containers for carbonated beverages to industrial fibers and films. The foil is glued on top of a rigid carrier, which is typically a *Si* wafer. After the production process the foil is delaminated from the carrier and the flexible foil with the circuits is left over.

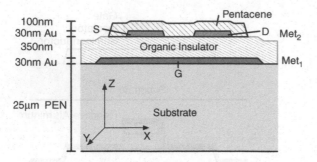

Fig. 2.11 Cross-sectional schematic view of a transistor architecture in the pentacene transistor technology. The *S*, *D*, and *G* contacts correspond to the source, drain, and gate of the transistor

The technology is extensively elaborated in Gelinck et al. (2004, 2005). A cross-sectional view of the technology is presented in Fig. 2.11. The thin-film transistors are built on top of the PEN substrate through a process which consists of six different layers. First a 30 nm *Au* metal layer Met_1 is deposited and patterned by photolithography. This layer serves as the bottom-gate of the transistor. Next an organic insulator is applied from solution. The thickness t_{ox} of this insulator is 350 nm. The term t_{ox} is actually not correct since the insulator is not an oxide, as in standard CMOS, although the term is used in this dissertation to emphasize the correspondence between an organic field-effect transistor and a standard MOS transistor. Subsequently another patterned *Au* layer Met_2 is deposited which provides the source and drain contacts of the transistor. The pentacene layer is formed by the solution-based spin-coating of a pentacene precursor, followed by a precursor-to-pentacene conversion step. This layer forms the channel of the transistor. These four layers enable the fabrication of a transistor.

The top view of the architecture of a typical transistor is shown in Fig. 2.12. The minimal length used for the transistors is 5 μm. This is not a hard limit but rather a safety limit. A smaller transistor length makes faster transistors but the reliability goes down due to a higher chance for malfunctioning transistors. The minimal width of the source and drain fingers is 5 μm for the same reason of reliability. Typically there is a 5−10 μm overlap of the pentacene layer beyond the source and drain fingers. This is to ensure a good alignment and a good contact between the pentacene and the electrodes. Finally, there is a gate overlap of 5−10 μm over the pentacene. This gate overlap is needed for reasons of alignment. Moreover the pentacene overlap sometimes causes leakage paths in the transistor. In order to control and limit this leakage it is important that the gate overlaps the pentacene layer. The overlap of the gate causes large overlap capacitances with both the source and the drain contact. These overlap capacitances ruin the speed of the circuits, even though this transistor architecture is required to ensure reliable behavior.

The measured $V_G - I_D$ curve a transistor with a 140 μm/5 μm W/L is shown in Fig. 2.13, both on a linear and a logarithmic scale. In the logarithmic graph the subthreshold slope can be derived. It amounts to around 2 V/dec. The measured

Fig. 2.12 Schematic top view of a 500 μm/5 μm pentacene bottom-gate transistor layout with 5 fingers. There is a 5 μm pentacene overlap over the drain and source contacts, and a 5 μm gate overlap over the pentacene

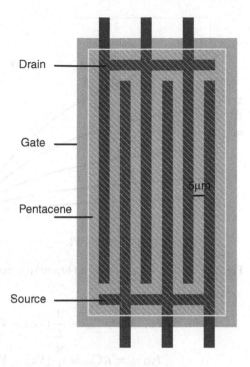

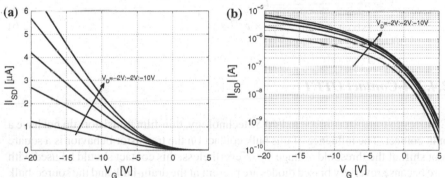

Fig. 2.13 The $V_G - I_D$ curves of a 140 μm/5 μm transistor, (**a**) on a linear scale, and (**b**) on a logarithmic scale

$V_D - I_D$ curve of the same transistor is shown in Fig. 2.14. The Early voltage can be derived and amounts to 5 V/μm. The transistor follows the typical MOS transistor behavior, given in Eq. (2.4) for the saturation regime, (2.5) for the linear regime, and (2.6) for the subthreshold regime:

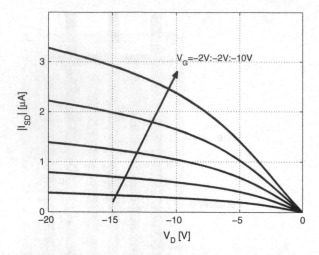

Fig. 2.14 The $V_D - I_D$ curves of a 140 μm/5 μm transistor on a linear scale

$$I_{SD,sat} = \mu C_{ox} \times \frac{W}{L}(V_{SG} - V_T)^2 \times \left(1 + \frac{V_{SD}}{V_E \times L}\right) \tag{2.4}$$

$$I_{SD,lin} = \mu C_{ox} \times \frac{W}{L}(V_{SG} - V_T) \times V_{SD} \tag{2.5}$$

$$I_{SD,sub} = I_{d,0} \times \frac{W}{L} \times e^{\frac{V_{SG}}{nkT/q}} \tag{2.6}$$

2.4.2 4-Contact OTFT

Contrary to transistors in a standard Si technology, thin-film transistors do not have a bulk contact. The influence of this bulk contact on the transistor behavior is a square root shift of the threshold voltage V_T. Nevertheless, this contact should be used with care because reversely biased diodes are present at the drain-bulk and the source-bulk interfaces. Furthermore the bulk contact of a Si transistor is sensitive to substrate noise that is caused by neighboring transistors. This is of course a consequence of the conductivity of the substrate.

Although thin-film transistors do not have an intrinsic bulk contact, there is the spatial opportunity to create a fourth contact on top of the transistor that can influence the electric field in the TFT channel. In this section, this technique is elaborated and the specific applicability of this technique for analog circuit applications is investigated.

The structure of an organic thin-film transistor with a backgate is presented in Fig. 2.15. On top of the original transistor presented in Sect. 2.4.1 two additional layers are deposited. First a 1.4 μm thick organic insulator is applied through spin-

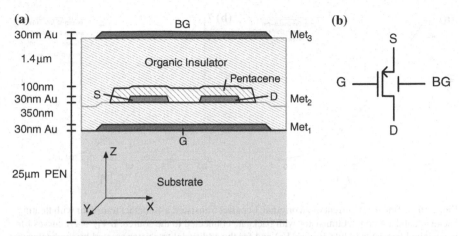

Fig. 2.15 a Cross-sectional schematic view of a transistor architecture with backgate in the dual-gate pentacene transistor technology. The *S*, *D*, *G*, and *BG* contacts correspond to the source, drain, gate, and backgate of the transistor. **b** The schematic symbol that represents the 4-contact transistor

coating. Finally a third patterned *Au* layer *Met₃* is deposited. The latter is used for interconnect and it is also applied as a backgate for the transistor. The influence of this contact on the transistor behavior is comparable to, yet different from, the bulk contact in standard CMOS transistors. The backgate has the same area as the gate of the transistor. Sometimes this backgate is also called a top-gate according to its physical location relative to the transistor channel. However, the functionality of the backgate is different from the functionality of the gate and in terms of transistor behavior it is slightly inferior to the gate. *Backgate* is therefore a better and more intuitive name for this contact.

2.4.2.1 Backgate Functional Model

The presence of the backgate on top of the original transistor influences the current characteristics of the original transistor. The difference between a transistor without a backgate, a transistor with a floating backgate, and a transistor with a backgate connected to the source is visualized in Fig. 2.16. It is clearly visible that the threshold voltage of the transistor with the floating backgate has shifted with more or less 5 V. The V_T value is unpredictable since it depends on the charge captured on the floating backgate. When no charge is captured on the backgate, i.e., when it is connected to the source, the V_T remains unchanged compared to the V_T of the original 3-contact transistor. The mobility of the transistor on the other hand is not affected by the presence of the backgate.

The physical location of the backgate creates a second transistor on top of the original 3-contact transistor. This second thin-film transistor is built with the backgate

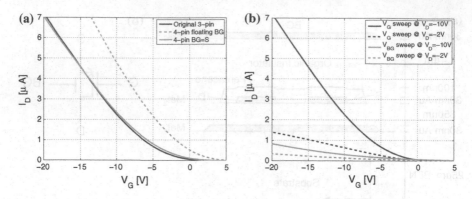

Fig. 2.16 **a** Transistor current of an original 3-contact transistor, a 4-contact transistor with floating backgate, and a 4-contact transistor with backgate connected to the source. **b** $V_G - I_D$ curves for the original transistor (while $V_{BG} = V_S$) and for the additional transistor created by the backgate (while $V_G = V_S$). The *straight lines* represent a $V_D = -10$ V the *dashed lines* a $V_D = -2$ V

contact and with the source and the drain. So the intuitively expected behavior of the backgate is a parallel transistor behavior. The second transistor has the same W and L dimensions, yet the insulator capacitance C_{ox} is about 5 times smaller due to the thickness of the second insulator layer, as can be seen in Fig. 2.15. The measured $V_G - I_D$ curves of both transistors are visualized in Fig. 2.16. The measurements of the gate transistor (gate, drain, source) are performed while the backgate is connected to the source. Those of the backgate transistor (backgate, drain, source) are measured while the gate is connected to the source. It can be seen that for a V_D of -2 V the ratio between the currents more or less equals 5 as expected. However, for a V_D of -10 V the ratio becomes larger and amounts to $\sim$7, likely caused by a shift of the V_T. This supports the idea that other effects are present, which is an acceptable idea since the electric fields of both transistors have a mutual influence on one another.

A better insight into the functionality of the backgate is acquired when the $V_G - I_D$ curves are plotted for various values of the backgate voltage V_{BG}. The measured transistor curves are plotted in Fig. 2.17. The black line corresponds with the characteristics of a transistor with a W/L of 140 μm/5 μm with the backgate connected to the source, which is the ground level. The gray curves correspond to the same transistor with a gradually varying backgate voltage V_{BG}. The leftmost curve corresponds with a V_{BG} of 10 V and the rightmost curve with a V_{BG} of -10 V. The curves in between correspond to intermediate backgate voltages with a 2 V step in between consecutive curves. It is remarkable that the curves shift in a very linear way. The conclusion that can be drawn from this graph is that a linear V_T shift occurs which is caused by the applied backgate voltage, i.e., the V_T becomes a linear function of the source-backgate voltage V_{SBG}. This function is written in Eq. (2.7).

$$V_T = V_{T,0} - \xi \times V_{\text{SBG}} \tag{2.7}$$

$$\xi = \frac{C_{\text{ox,t}}}{C_{\text{ox,b}}} = \frac{t_{\text{ox,b}}}{t_{\text{ox,t}}} \approx 250\,\text{mV/V}, \tag{2.8}$$

where $V_{T,0}$ is the V_T of the original 3-contact transistor and ξ is a constant that represents the slope of the dependency. This value ξ can be extracted from the measurements presented in Fig. 2.17 and amounts to 350 mV/V. The theoretical value of ξ corresponds with the ratio of the insulator capacitance of the bottom transistor $C_{\text{ox,b}}$ and the insulator capacitance of the top transistor $C_{\text{ox,t}}$ (Gelinck et al. 2005). It is deducted in Eq. (2.8). The calculated value does not match very well with the measured value. This can be caused by a change in the process after the publication of Gelinck et al. (2005). The measured value will be further employed in this work. Furthermore, the value of ξ is different for positive and negative voltages, according to Maddalena et al. (2008). Nevertheless, the measurements showed that in the applied technology these values are approximately identical hence they permit to use only the one measured value for ξ in this work. The V_T shift can in a simplified way be explained by the electric field in the Z-direction that is superposed on the electric field that is already present in the original transistor. This superposed electric field is linear with the applied voltage, just like in a capacitor. This electric field hampers the accumulation of charge carriers and must additionally be neutralized, in the case of a positive backgate bias. On the other hand, in the case of a negative backgate bias, the required electric field to enable accumulation is already partially present and the V_T is then shifted to a more positive voltage.

The inclusion of the effect of the backgate on V_T in the simplified formula for the transistor current in saturation Eq. (2.4) is derived in Eq. (2.9):

$$I_{SD,\text{sat}} = \mu C_{\text{ox}} \times \frac{W}{L}(V_{\text{SG}} + \xi \times V_{\text{SBG}} - V_{T,0})^2 \tag{2.9}$$

This equation demonstrates that the current of the transistor is determined by the sum of the gate voltage and the weighted backgate voltage. Both contacts have an identical influence on the transistor current, albeit with a different weight. The ratio of the effect of gate and backgate on the transistor current is given by $1/\xi$ or $3/1$.

The effect of the backgate on the transistor behavior has in the previous paragraphs been investigated and derived only in the $V_G - I_D$ plane, hence at a fixed V_D. In order to generalize the description of the 4-contact transistor the effect of the backgate should also be discussed in the $V_D - I_D$ plane, i.e., with a fixed V_G. Figure 2.18 shows the $V_D - I_D$ curves of a 140 μm/5 μm transistor without a backgate and a 140 μm/5 μm transistor with a backgate connected to the source. The curves of the 3-contact and the 4-contact transistors show a certain correspondence, yet there is a very important difference between them, namely in the slope of the curves in the saturation regime. The transistor with the backgate has a higher output impedance r_{sd}. The reason for this improved behavior is a change of the Early voltage due to the changes in the physical structure of the transistor. The Early voltage can be graphically deduced from the figure and amounts to approximately 5 V/μm for the

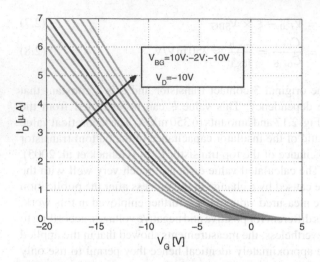

Fig. 2.17 Measured $V_G - I_D$ characteristics of a 140 μm/5 μm with a $V_m at x BG$ that varies between −10 and 10 V. The *black curve* corresponds to the case where $V_{BG} = 0$ V. The measurements are performed with a $V_S = 0$ V and a $V_D = -10$ V

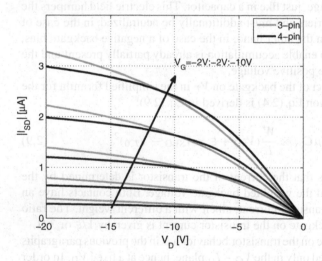

Fig. 2.18 Measured $V_D - I_D$ characteristics of a 140 μm/5 μm transistor without a backgate and a 140 μm/5 μm transistor with a backgate connected to the ground. The V_G varies between −2 and −10 V . The curves of the transistor with the backgate (4-contact) have a larger output resistance in the saturation region than the 3-contact transistor curves

3-contact transistor and approximately 12 V/μm for the 4-contact transistor. The output resistance r_{sd} of the transistor is given by Eq. (2.10):

$$r_{sd} = \frac{V_E \times L}{I_{SD}} \qquad (2.10)$$

hence a transistor with a fixed backgate has an output resistance which is about 2.4 times higher than a transistor without a backgate. When the backgate of the transistor is connected to the gate also the current I_{SD} increases with a factor 1.8 in (2.10) and then the output impedance increases with a factor 1.33. The formula for the transistor current in saturation in Eq. (2.9) leaves the $V_D - I_D$ behavior out of account. The more detailed formula with the $V_D - I_D$ behavior included is written in Eq. (2.11):

$$I_{SD,sat} = \mu C_{ox} \times \frac{W}{L}(V_{SG} + \xi \times V_{SBG} - V_{T,0})^2 \times \left(1 + \frac{V_{SD}}{V_E' \times L}\right) \qquad (2.11)$$

$$V_E' = 2.4 \times V_E \qquad (2.12)$$

The model derived in this section takes into account the behavior of a transistor with a backgate contact in the saturation region. It is further used in this dissertation to model the transistors with a backgate, wherever they are used. More detailed physical information on the backgate and, in general, on dual-gate transistors can be found in Spijkman et al. (2011).

2.4.2.2 Backgate Versus Bulk

The backgate of the organic thin-film transistor has, up to a certain level, a similar behavior to that of the bulk contact of a MOS transistor. It is worth the effort to make a little comparative study between both contacts and their influence on the transistor characteristics. Equations (2.13)–(2.15) represent the $V_{T,MOS}$ and the transistor current I_{MOS} of a PMOS transistor (Sansen 2008).

$$V_{T,MOS} = V_{T,0} + \gamma \left(\sqrt{|2\phi_F| + V_{SB}} - \sqrt{|2\phi_F|}\right) \qquad (2.13)$$

$$K_p' = \frac{\mu C_{ox}}{2\gamma} \times \sqrt{|2\phi_F| + V_{SB}} \qquad (2.14)$$

$$I_{MOS} = K_p' \times \frac{W}{L}(V_{SG} - V_{T,MOS})^2 \qquad (2.15)$$

In these equations, $V_{T,0}$ is the initial threshold value, γ a technology-dependent factor, ϕ_F the Fermi level, and K_p' is the transistor constant in saturation. The influence of the bulk on the transistor behavior is rather complex. A more negative bulk voltage V_B decreases the threshold voltage $V_{T,MOS}$ and accordingly decreases the transistor current. The influence on V_T follows a square root relationship whereas the influence of V_{SB} on the transistor current is more complex. The explanation for the behavior of the bulk contact is found in the increase of the depletion layer under the transistor channel at the bulk-source interface for an increased V_{SB}. The influence of the source-backgate voltage V_{SBG} on the V_T and the transistor current in saturation has been described in Eqs. (2.7) and (2.11). The striking difference between the bulk contact in a PMOS transistor and the backgate contact in a p-type organic thin-film transistor (POTFT) is that the influence of the backgate on the POTFT current is opposite to that of the bulk on the PMOS current. This means that new circuit techniques can be found and applied that adopt this behavior and improve the transistor behavior. Moreover, since the backgate contact is fully disconnected from the transistor channel, no attention must be spent to leakage currents or to diodes that have to be kept reversely biased, like with the bulk contact in Si CMOS. Furthermore, the nice linearity of the influence of the backgate on the V_T for both positive and negative V_{SBG} makes the backgate in OTFTs a very useful contact for the design of both digital and analog circuits.

2.4.3 Features of the 4-contact OTFT

The presence of the backgate in the 4-contact transistor enables to change the threshold voltage of the transistor. The behavior of the modified transistor is investigated and a functional model that includes the backgate contact has been presented in Sect. 2.4.2.1. In this Section the 4-contact transistor is investigated in a more application-oriented way that focuses on the design of analog circuits in order to optimally exert all its available features.

2.4.3.1 Threshold Voltage Tuning

It is demonstrated in Sect. 2.4.2.1 that V_T is proportional to V_{SBG} in a broad voltage range. Therefore, the backgate contact can be useful in a transistor of which the V_T or $V_{SG} - V_T$ should be tuned. This is the basic principle of a very interesting technique which is related to the DC levels of the gate, drain, and source contacts of the transistor: indirectly the backgate contact enables a degree of freedom in the choice of the DC levels of the transistor. Typically in the 3-contact transistor the drain current is directly determined by V_{SG} and the V_{SD}, hence there are two degrees of freedom. This means of course that when the current in the transistor is kept constant to a certain level the V_{SG} and V_{SD} can no more be both independently chosen. This can in the unipolar transistor technology lead to troubles with unequal DC bias levels in analog circuits such as differential amplifiers. The 4-contact transistor enables an additional degree of freedom through the backgate contact. This backgate is specifically useful in combination with the gate. From Eq. (2.11) it can easily be derived that a change of the gate voltage ΔV_G is counterbalanced in such a way that the current is kept constant. The V_{SBG} required for this operation is derived in Eq. (2.16):

$$V_{SBG} = -\frac{\Delta V_G}{\xi} \qquad (2.16)$$

The interpretation of this formula is that in order to decrease the DC level of the gate contact of the transistor with 1 V the backgate DC level must be increased with about 3 V. The linearity of the V_T shift is valid for a very large interval, reaching over several tens of volts, and in this way the backgate can enable the free choice of the DC level of the gate contact without changing I_{SD} or V_{SD} of the transistor. This technique is applied for DC connecting consecutive differential amplifier stages in Sect. 3.3.3.

2.4.3.2 Gate-Backgate Connected Transistors

A second and probably more powerful technique that fully turns the backgate to advantage is obtained by connecting the backgate contact with the gate contact. The technique in this work is also referred to with the term *backgate steering*. In this technique the V_T of the transistor is shifted during operation by the gate voltage itself. The result of this V_T shift on the $V_G - I_D$ curve is plotted in Fig. 2.19a. The gray curves correspond to the measured $V_G - I_D$ curves of the 4-contact 140 μm/5 μm transistor with a fixed backgate voltage that is stepwise varied between 10 and −10 V. Each of these lines include one dot where V_G equals V_{BG}. All those dots also correspond to the curve of the transistor with its gate and backgate connected. As such the curve of the gate-backgate connected transistor is graphically derived. The black curve corresponds to the measured $V_G - I_D$ curve of this transistor and logically coincides with the graphically derived curve. The acquired transistor curve is clearly steeper than the original curves, i.e., more current flows through the same transistor under the same V_{SG} and V_{SD} and moreover a higher transconductance g_m is obtained. This behavior can be derived from Eq. (2.9) and the result is presented in Eq. (2.17).

$$I_{SD,sat} = \mu C_{ox} \times \frac{W}{L}(V_{SG} + \xi \times V_{SBG} - V_{T,0})^2 \quad \text{(with } V_{SBG} - V_{SG})$$

$$\Downarrow$$

$$I_{SD,sat} = (1 + \xi)^2 \times \mu C_{ox} \times \frac{W}{L}\left(V_{SG} - V'_T\right)^2 \quad \text{(where } V'_T = \frac{V_{T,0}}{1 + \xi}) \quad (2.17)$$

The striking conclusion from this equation is that the transistor current still follows the quadratic current law in saturation, which facilitates the modeling of this behavior. According to the measured value of 350 mV/V for ξ, an increase of the transistor constant K'_p with 82 % is experienced. Furthermore, the V_T of the transistor is slightly moved by this technique. The increase of the current is logically passed on to the transconductance, where an identical increase is experienced when the gate overdrive $V_{SG} - V_T$ is kept constant, as can be seen in Eq. (2.18):

$$g_m = \frac{2I_{SD}}{V_{SG} - V'_T} = (1 + \xi)^2 \times g_{m,0} \qquad (2.18)$$

where $g_{m,0}$ is the original g_m value of the 4-contact transistor with fixed backgate. It must be noted that this comparison is interpreted for identical gate overdrive voltages,

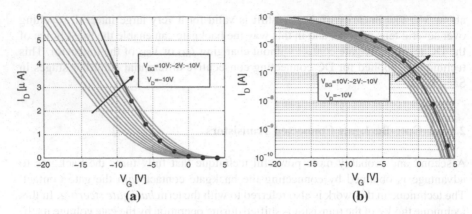

Fig. 2.19 Graphical derivation of the transistor characteristics of the gate-backgate connected 4-contact OTFT. The *gray lines* correspond to the 4-contact OTFT with fixed backgate voltages that reach from -10 to $10\,V$. The *black dots* represent the unique point on every curve in which the gate voltage equals the backgate voltage. The *black line* corresponds to the measured characteristics of the gate-backgate connected 4-contact OTFT. The measurements are performed with a $V_D = -10\,V$

i.e., $V_{SG} - V_T'$ of the gate-backgate connected transistor must be identical to the $V_{SG} - V_{T,0}$ of the original 4-contact transistor.

In Fig. 2.19b, the same $V_G - I_D$ curves are plotted yet on a logarithmic scale, which enables to investigate the transistor behavior in the subthreshold region. Through the same principle as in the saturation region the transistor curve of the gate-backgate connected transistor becomes steeper, i.e., the subthreshold slope is higher than when the backgate voltage is kept constant. The subthreshold slope can be graphically esti-mated from the figure and decreases from 2 to $1.2\,V/dec$. A smaller subthreshold slope improves the 'off' behavior of the transistor, which is beneficial for digital circuits, where a high ratio between on current and off current is desired. For the design of analog circuit design the subthreshold slope does not have important con-sequences, since transistors are almost never intentionally biased in the subthreshold region.

In Fig. 2.20 the behavior of the gate-backgate connected transistor is highlighted in the $V_D - I_D$ plane. The gate-backgate connected transistor draws a clearly higher current than the transistor with the backgate connected to the source contact. The increase of the current is explained by Eq. (2.17) and depends on the V_{SG} rather than on V_{SD}. It is also visible that the output resistance r_{sd} decreases for the same reason. However, a more important conclusion from this graph is that the ratio of the output conductance and the transistor current remains unchanged when this technique is applied. This is easily demonstrated by the two accidentally closely lying curves, i.e., the gray curve for a $V_G = -8\,V$ and the black curve for a $V_G = -6\,V$ that are parallel in the saturation region.

The physical presence of the backgate on top of the transistor also has its influence on the capacitances that are experienced at the contacts. The backgate capacitance

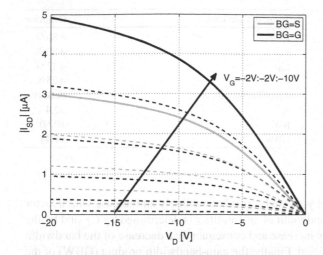

Fig. 2.20 Measured $V_D - I_D$ characteristics of a 140 μm/5 μm gate-backgate connected transistor and a 140 μm/5 μm transistor with a backgate connected to the ground. The V_G varies between -2 and -10 V

C_{BG} is one-fourth of the gate capacitance C_G according to the thicknesses of the respective insulator layers. This means that the parasitic capacitances on the contacts in the gate-backgate connected transistor are increased with 25 % which typically slows down the speed of analog circuits. However, the current through the transistor is increased with 80 %, hence both effects result in a favorable balance for the speed in the transistor.

The connection of the backgate to the gate influences the behavioral transistor parameters and accordingly it will influence the specifications of analog circuits. An increase of the transistor current and of the g_m has been demonstrated. The ratio g_m/I_D, however, remains unchanged. The same reasoning is valid for the impedance r_{sd} of the transistor. It decreases proportionally with the transistor current. Another effect also influences the value of r_{sd}. The presence of the backgate, independent of the biasing scheme, increases the V_E with a factor 2.4. The r_{sd} is determined by this value and by the current I; hence it increases with a factor 1.33. The only possible drawback of this topology is found in the increase of the gate capacitance with 25 %. The effect of applying this technique in a single-stage analog amplifier on the gain, the gain-bandwidth product, and the bandwidth of that amplifier can be predicted through Eqs. (2.19)–(2.21):

$$A = g_m \times r_{sd} \tag{2.19}$$

$$BW = \frac{1}{2\pi \times r_{sd} \times C_L} \tag{2.20}$$

$$GBW = \frac{g_m}{2\pi \times C_L} \tag{2.21}$$

Table 2.1 Overview of the consequences of gate-backgate steering on the transistor behavior and on the specifications of an analog amplifier

Transistor parameter	Behavior
I_D	+80 %
g_m	+80 %
r_{sd}	+33 %
C_G	+25 %
Amplifier specification	Behavior
A	x2.4
BW	x0.60
GBW	x1.44

The gain (A) is determined by g_m and r_{sd} and accordingly an increase with a factor 2.4 can be expected. The bandwidth of an amplifier is determined by C_L and by the r_{sd}. Both parameters slightly increase and consequently a decrease of the bandwidth (BW) of about 40 % is predicted. Finally, the gain-bandwidth product (GBW) of the amplifier is the product of the two previous results and an overall GBW increase with factor 1.44 is expected. The expected influence of the gate-backgate steering technique on the gain, the gain-bandwidth product and the bandwidth of an amplifier are summarized in Table 2.1.

2.4.3.3 Gate-Backgate Connected Diodes

The third and last example that demonstrates the performance increase which is obtained by the use of backgates is a diode-connected transistor. Diode-connected transistors are employed several times in this work, especially in Chap. 6 for the passive implementation of switches in DC–DC converters.

The application of a backgate which is connected to the gate and the drain increases the I_{on}/I_{off} ratio. This is made clear in Fig. 2.21 where the measured curves of a 3-contact and a 4-contact diode-connected transistor are shown. While the diodes are biased on, the current of the 4-contact diode is steeper and higher than the current in the 3-contact variant, whereas the current in the 4-contact diode becomes lower than in the 3-contact diode while the diodes are biased off. The result of the improvement is an increase of the I_{on}/I_{off} ratio with a factor 2.5 calculated with the measured diode currents at ±20 V.

2.5 Non-Ideal Behavior

The transistor model that has been discussed in Sects. 2.4.1 and 2.4.2 for the 3-contact and the 4-contact transistor respectively is a simplified model. In practice on top of this behavior a lot of secondary but often important effects come to the surface. They are mostly undesired and therefore classified as non-ideal behavior. The production

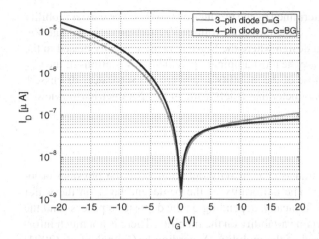

Fig. 2.21 Measured characteristics of a $140\,\mu$m/$5\,\mu$m original 3-contact transistor diode and a $140\,\mu$m/$5\,\mu$m 4-contact transistor with the backgate connected to the gate

of organic electronics circuits is the subject of a lot of technological research all over the world. The focus of this research is to make the technology more performant and to reduce non-ideal behavior that is present. This non-ideal behavior can be caused during production, e.g., by a variation of the thickness of a layer, or it can be associated to the employed materials, e.g., the sensitivity of pentacene to the presence of H_2O in the ambient environment which results in degradation. It is important for the circuit designer of either digital or analog circuits to be aware of those behaviors so that they can be overcome at the circuit level. In this section, the non ideal behavior of organic pentacene-based thin-film transistors on foil is explored.

2.5.1 Process-Induced Influences

The cross-sectional view of a transistor in the applied organic electronics technology is shown in Fig. 2.15. The transistor is built with six different layers of which three metal layers. The transistor behavior is in all possible ways determined by these layers: The absolute or relative thickness of the layers, the temperature of the wafer during a deposition step, the temperature of the material that is deposited during the deposition, the composition of the environment in the deposition chamber, e.g., vacuum or ambient, the speed at which the layers are deposited. The list of parameters that influence the transistor behavior is endless. A profound examination of the effect of all those parameters is beyond the scope of this work. It is, however, very interesting and important to have a look at the influence of all these production parameters together on the behavioral transistor parameters. The current of a transistor biased in the saturation regime is written in Eq. (2.4). In this formula four parameters are present which are directly determined by the production process. The carrier mobility μ, the insulator capacitance C_{ox}, the threshold voltage V_T, and the early voltage V_E. In the scope of analog circuit design where the transistors are typically biased in the saturation regime, V_T and μ are the most dominant parameters. Variations of the C_{ox} have an identical effect to those of the mobility μ and are not separately discussed. The V_E is a less dominant parameter that is left aside in this parametric overview.

2.5.1.1 Mobility

The carrier mobility of the transistor is heavily determined by the physical properties of the pentacene layer. The uniformity of the pentacene layer on the wafer must therefore be optimized. Typically, depending on the deposition process and the deposition tool, there is a certain variability on the mobility. There is not much information available on the spread of the mobility. According to Gelinck et al. (2004) the on-wafer mobility has a mean value m_μ of $0.02\,\mathrm{cm^2/Vs}$ and a standard deviation σ_μ of $0.04\,\mathrm{cm^2/Vs}$. However, according to Myny et al. (2009) nowadays the mean value m_μ of the mobility in the technology of the same producer has increased up to $\sim 0.15\,\mathrm{cm^2/Vs}$. Information on the standard deviation is not given there but it is assumed reasonable that the relative deviation given by σ_μ/m_μ has not increased during those 5 years of improvement. Consequently the spread of the mobility is not expected to be a limiting factor of the circuit complexity. Furthermore, the most important factor for circuit design is the parameter matching between adjacent transistors. The short distance standard deviation $\sigma_{\mu,adj}$ is in the applied technologies lower than σ_μ of the full wafer.

The constant search for improvement often implies small changes in the production process which is optimized for flexible display applications. There is often a time gap of several months between tape-outs, it can occur that the mobility has changed compared to its value in the previous tape-out. Most often a slight increase, but possibly a slight decrease of the mobility is then expected and therefore these inter-tape-out variations only result in an increase or a decrease of the bandwidth of the circuits.

2.5.1.2 Threshold Voltage

The threshold voltage of a transistor is a behavioral constant that can be graphically derived as the intersection of the tangent line of the square root of the transistor current and the X-axis. It defines a threshold between the on-state and the off-state. The threshold voltage is heavily determined by the electric field that is present in the transistor channel and therefore to small fixed charges present near the transistor channel. The net amount of charge that is trapped near the channel is a very uncertain value which is very sensitive to almost any parameter in the production process.

The threshold voltage on a wafer is examined in Gelinck et al. (2004). The mean threshold voltage m_{V_T} is $\sim 0.5\,\mathrm{V}$ and the standard deviation σ_{V_T} amounts to $0.25\,\mathrm{V}$.

These data are old but nevertheless from this result two conclusions can be drawn. In the first place it is uncertain whether the value of V_T is positive or negative. This means that when a transistor is biased with a V_{SG} of 0 V, i.e., in the zero-V_{GS} configuration, it is uncertain whether the transistor is biased on or off. In the second place it must be clarified that the relative variation of the V_T is not of importance, but rather the absolute σ_{V_T} which is relatively limited over the area of the wafer.

The main risk related to V_T variation is the variation between tape-outs. The constant search for improvement often implies small changes in the production process that is optimized for flexible display applications. The high sensitivity to a slight change in the process can cause significant V_T changes in between tape-outs. When biased near V_T, the transistor is very nonlinear element; therefore, it is very unreliable to bias the V_{SG} of a transistor close to its threshold voltage. Moreover, unlike for the mobility, it is not logically predictable whether the V_T change will occur in a positive or a negative direction. This problem is further extensively discussed in Sect. 3.2.3.

2.5.2 Environmental Influences

A wafer with organic electronics circuits that is placed in the ambient environment is exposed to a large set of environmental parameters, e.g., temperature, light, air, consisting of N_2, O_2, CO_2 and H_2O, etc. The behavior of pentacene is susceptible to changes of some of those parameters. Therefore, besides the effects which occur during the production of the wafers there are also undesired effects which occur after the production, either while being shelved or at runtime. This means that the transistor behavior will change over time as a function of external parameters. This dependency can be advantageous for sensor design. However, as will be discussed in Chap. 5, an unaffected transistor behavior is desired for all other circuits. A profound examination of the effect of all the environmental parameters is beyond the scope of this work. Only a general overview of the most important effects is presented in this section.

2.5.2.1 Temperature

The current through a pentacene transistor depends on the environmental temperature. This is explained by a temperature-dependant mobility μ and threshold voltage V_T.

In Wilke and Gobel (2008), the carrier mobility of pentacene is investigated in the 300–400 K temperature range. They report an increasing mobility with increasing temperature up to a certain critical temperature T_C. Above that temperature the mobility decreases irreversibly with increasing temperature. According to Guo et al. (2007) the threshold voltage is found to decrease with the increasing temperature. That paper also suggests that defects in the materials limit the mobility and that it is possible to obtain significantly higher mobilities when those defects in the materials are substantially removed.

2.5.2.2 Bias Stress

The presence of an applied electrical field over time, such as applied during operation, tends to influence the behavior of organic transistors. This is called bias stress. Bias stress in organic semiconductors follows from carrier trap states at the dielectric–semiconductor interface or in the dielectric that compensate for the applied electric field (Ng et al. 2007; Genoe et al. 2004). The compensation for the electric field is experienced through a shift of the threshold voltage of the transistor. The consequences of bias stress depend on the way the stress is applied, i.e., whether the V_{SG} is negative or positive. Bias stress at negative V_{SG} (with the pentacene transistor in the off-state or mild electron accumulation) is faster and more degrading than bias stress at positive V_{SG} (transistor in on-state with hole accumulation) (Genoe et al. 2004). Furthermore, bias stress is proportional to the applied V_{SG} and increases with increasing bias time. Some papers report a slow recovery, with a time constant of 2–4 days, of the threshold voltage from bias stress in the unstressed state (Street et al. 2006). Bias stress can be exacerbated by the presence of water and oxygen molecules at the semiconductor, as well as by light (Pernstich et al. 2006; Diallo et al. 2007; Debucquoy et al. 2007; Wrachien et al. 2010). The effect of oxygen is explained by chemical reaction with the pentacene molecule, according to Refs. Pannemann et al. (2004, 2005). Other Refs. Vollmer et al. (2006); Jurchescu et al. (2005) explain the behavior by diffusion of O_2 in the pentacene layer. Besides the shift of V_T also a mobility degradation is reported when transistors are biased in ambient environment containing O_2 and H_2O for long time (Street et al. 2006).

A bias stress measurement has been performed to estimate the consequences in the applied organic electronics technology. A transistor with a W/L of 7000 μm/5 μm has been stressed for 24 h with its gate and drain contacts biased to -15 V under ambient conditions, i.e., with normal exposure to O_2, H_2O, light ,and temperature fluctuations. The measurement results are presented in Fig. 2.22. A clear shift of the transistor curve is visible that corresponds with a V_T shift of 5 V. A degradation of the mobility is not present. This can be due to the short stress time in the experiment. The presented measurement has only been performed on a single transistor and the statistical value of the numerical result is therefore low. Nevertheless, the presented result is strong enough to conclude that V_T rather than μ is the transistor parameter that is by far most affected by bias stress.

2.5.2.3 Unstressed Degradation

The presence of H_2O and O_2 in the ambient environment and the presence of light are known to influence the behavior of a pentacene transistor. In Pannemann et al. (2004), a mobility degradation of a decade per quarter of a year is experienced with

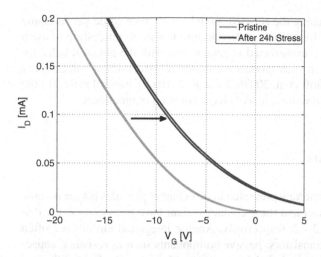

Fig. 2.22 Measured $V_G - I_D$ characteristics of a pristine 7, 000 μm/5 μm transistor and after a 24 h bias stress under ambient conditions, i.e., with ambient exposure to H_2O and O_2 in the air, and to light and temperature fluctuations (Street et al. 2006). The gate contact and the drain contact have been biased with -15 V. The effect is mostly a shift of the V_T

unprotected pentacene transistors that were shelved in dark under ambient conditions. Their measurements also revealed a 12.8 V shift of V_T after nine months shelving. This shift is reversible when the transistors are shelved in a vacuum environment. Encapsulation of the circuits directly after the production with a hydrophobic material is introduced in Pannemann et al. (2005) as a solution that prevents the highly sensitive pentacene layer from degradation. Contradictory to Pannemann et al. (2004), according to Kagan et al. (2005), the degradation is negligible when the pentacene circuits are only shelved in ambient without being electrically biased.

It can be assumed that storing organic circuits in ambient environment may have certain consequences on the transistor behavior due to the presence of H_2O and O_2. Therefore, it is more safe to store organic transistors and circuits in nitrogen (N_2) before the measurements are performed. The risk for unstressed degradation is then extensively reduced. Another way to reduce the sensitivity to degradation is to encapsulate the transistors with a thick layer of material during production.

2.6 Organic Transistor Modeling

Until this point, but also further on in this dissertation, a straightforward MOS transistor model is adopted for fitting the organic thin-film transistor behavior. Although TFTs show a certain behavioral correspondence with standard CMOS transistors, their small-scale physical behavior is totally different from that in a CMOS transistor, as explained in Sect. 2.2. At the moment the complexity of organic circuits is still small enough to keep using the adopted model for another while yet for some applications and especially for a better understanding it is more interesting to use a physical OTFT model that is constructed in accordance with the physical behavior of the OTFT. At the moment such models are under construction and nice results have been presented (Oberhoff et al. 2007b; Li et al. 2010a,b). Nevertheless, for the simulations in this dissertation those models have not been implemented.

2.7 Other Components

The organic electronics technology presented in this chapter provides p-type organic transistors. Both the 3-contact and the 4-contact variant have been extensively discussed in Sects. 2.4.1 and 2.4.2, respectively. Analog integrated circuits are often built with more than only transistors: passive components such as resistors, capacitors, and inductors are often included in specific analog circuits. Especially in a unipolar transistor technology those components are a considerable contribution to the quality of a building block. In this section an overview is given of the feasibility of making and using passive components in the organic electronics technology.

2.7.1 Resistors

Resistors are often used in analog circuits, e.g., in RC filters. The behavior of an ideal resistor is described by Ohm's law, given in Eq. (2.22). A resistor built in an integrated circuit technology, though, is a non-ideal component whose behavior must only match with ideal resistive behavior in the specifications range of interest, e.g., the resistive behavior is only required in the frequency band for which the analog circuit is designed and up to a certain level of bias conditions that are typical for that analog circuit.

$$V = R \times I \quad (2.22)$$

The applied organic electronics technology does not provide specific layers for making resistors. Nevertheless, every material that is present on the wafer has its own intrinsic resistivity and can be employed as a resistor. The challenge is now to create resistors that have the desired resistive value, the desired linearity, and the desired speed. In this section an overview is given of the possible resistor architectures and their properties.

2.7.1.1 Metal Line Resistor

A first way to make a resistor is a long line of the gold layers that are available in the technology, as shown in Fig. 2.23a. The resistance per square of the gold layers

Table 2.2 Measured resistive values obtained from metal line resistors with a varying line width and line spacing

Width & spacing (μm)	1-metal R	3-metal R	1-metal R per mm^2 (kΩ/mm^2)
10	743 Ω	2.6 kΩ	4.7
5	3.4 kΩ	12.1 kΩ	21.6
2	23.4 kΩ	shorted	148

The measurements were performed on single-metal lines and on serial three-metal lines. All the measured resistors were built on a 0.4×0.4 mm^2 area.

is calculated from the thickness of the layer (30 nm) and from the resistivity of gold at room temperature (2.44 e–8 Ωm). It amounts to approximately 0.8 Ω/□. zigzag walking path with both a typical width and a metal spacing of 5 μm results in a resistance of approximately 20 kΩ/mm^2 per metal layer.

Metal line resistors with a width and a spacing of 10, 5, and 2 μm., respectively, on a 0.4×0.4 mm^2 area are measured and the measurements are summarized in Table 2.2. The measured resistance per area of 21.6 kΩ/mm^2 for the 5 μm line corresponds very well to the expected 20 kΩ/mm^2. Due to the low carrier mobility in the transistors the typical value of the node impedances in an organic circuit is much higher than the 10–100 kΩ range that is reached with metal line resistors. The resistance per area of a metal line resistor can be increased by decreasing the metal width and spacing. Halving both the width and the spacing increases the resistance per area with a factor 4 yet this also increases the risk of hard failures.

2.7.1.2 Pentacene Resistor

A resistive component can also be constructed with the pentacene layer. The layout of such a pentacene resistor is shown in Fig. 2.23b. The resistance of a pentacene thin-film resistor is unpredictable as the conductivity of pentacene depends on the electric field that is experienced, like in a pentacene thin-film transistor. The absence of a gate below the pentacene layer presumably results in a zero electric field in the pentacene and consequently is the resistance dependent on the position of the threshold voltage value and a large variation of the resistance, especially in between wafers and tape-outs is expected.

Measurements have been performed to estimate the resistance values of such transistors. They are summarized in Table 2.3, where very high resistances in the GΩ range are obtained as was expected. The mean resistance per square that is measured for this type of resistors is 19 GΩ/□. An eye-catching result is that the measured value of $R_{pent,3}$ differs from the measured value of $R_{pent,2}$, while an identical value is expected. This type of resistors is very unpredictable and it is preferable to choose other resistor types in this technology.

Table 2.3 Measured resistive values obtained from pentacene resistors

Name	Size	Sample 1 (MΩ)	Sample 2	Mean Value (MΩ)	R/Square (GΩ/□)
$R_{\mathrm{pent},1}$	140 μm/10 μm	1341	1439 MΩ	1390	19.5
$R_{\mathrm{pent},2}$	140 μm/25 μm	4223	4092 MΩ	4157	23
$R_{\mathrm{pent},3}$	280 μm/50 μm	2551	/	2551	14.5
$R_{\mathrm{pent},4}$	140 μm/50 μm	7434	6054 MΩ	6744	19

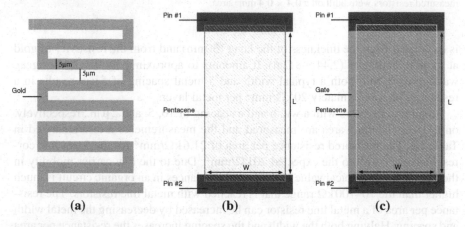

(a)   (b)   (c)

Fig. 2.23 a Schematic *top view* of a *metal line* resistor built in one of the available gold layers. **b** Schematic top view of the *pentacene* resistor architecture which is built up with pentacene and two gold contacts. **c** Schematic *top view* of the *linear-transistor* resistor architecture which is built up in a rectangular way without finger structures

2.7.1.3 Transistor in Linear Regime

Other conducting or semiconducting materials than gold and pentacene are not supplied in the applied organic electronics technology, yet there is another, probably more adequate way of making resistors. Transistors which are biased with a $V_{\mathrm{SG}} - V_T$ larger than their V_{SD} are in the linear regime. Their resistance is derived from Eqs. (2.5) and (2.22) and the result is given in Eq. (2.23):

$$R_{\mathrm{SD,lin}} = \frac{1}{\mu C_{\mathrm{ox}} \times \frac{W}{L}(V_{\mathrm{SG}} - V_T)} \tag{2.23}$$

This resistance is given by the technological constants μ and C_{ox}, by the physical sizes W and L, and by the gate overdrive $V_{\mathrm{SG}} - V_T$ that is typically large. With this type of resistors a relatively accurate resistance is obtained since variations of μ are only in the range of a few percents up to in the worst case a few tens of percents.

Since the working region as well as the dimensions of the transistors in this section are far from the more conventional transistor biasing (in saturation) and sizing which are used for fitting the transistor model it is safer to measure a set of these resistors

Table 2.4 Measured resistive values obtained from resistors built with transistors in the linear region

Name	Size	Sample 1 (MΩ)	Sample 2 (MΩ)	Mean Value (MΩ)	R/Square (GΩ/□)
$R_{\text{lin},1}$	140 μm/63 μm	127.5	130	128.75	286
$R_{\text{lin},2}$	280 μm/127 μm	113.5	157.5	135.5	298
$R_{\text{lin},3}$	560 μm/273 μm	120.5	124.5	122.5	251
$R_{\text{lin},4}$	140 μm/85 μm	175.5	177	176.25	290
$R_{\text{lin},5}$	140 μm/170 μm	333.5	398	365.75	301
$R_{\text{lin},6}$	140 μm/260 μm	607.5	518	562.75	303
$R_{\text{lin},7}$	140 μm/10 μm	11	/	11	154
$R_{\text{lin},8}$	140 μm/5 μm	6.9	6.8	6.85	192

A V_{SG} of 13.5 V was applied during the measurements. All R are built with the layout in Fig. 2.23, except for $R_{\text{lin},7} - R_{\text{lin},8}$ which are constructed like in 2.12.

than to extrapolate the model and rely on the simulated resistive values. A set of resistors with varying W and L have been measured at a V_{SG} of 13.5 V and with a V_{SD} close to 0 V. The results are listed in Table 2.4. The rectangular lay-out of the resistors without fingers is Fig. 2.23c.

The ratio of W/L for resistors $R_{\text{lin},1-3}$ is kept constant while the area is increased, hence an identical resistance is expected with a better matching for larger area according to Pelgrom's model (Pelgrom et al. 1989). It is not the aim of the author to provide an elaborate statistical study here yet it is visible that the mean resistances of $R_{\text{lin},1-3}$ are similar. The resistors $R_{\text{lin},4-6}$ all have a 140 μm width while the length is increased with a factor 1,2, and 3, respectively. This results in resistive values that are proportional to the length as expected. The mean resistance per square for resistors $R_{\text{lin},1-6}$ amounts to 288 MΩ/□. Finally, two transistors $R_{\text{lin},7-8}$ with the normal transistor lay-out (presented in Fig. 2.12) have been measured. Their resistance per square is notably lower and amounts to 170 MΩ/□. It is difficult to draw conclusions from those two but they give an estimation of the impedances that are present in an analog organic circuit and accordingly they provide a lower boundary for the resistance values of interest.

The advantage of these resistors is that the transistors are biased with a high gate overdrive and accordingly their speed is high, higher than the speed of analog circuits where transistors are biased with smaller gate overdrives, hence the bandwidth of these resistors should not limit the speed of a circuit in which they are used. The most important drawback of this type of resistors, on the other hand, is that the V_{SG} must be kept constant. Therefore, when a fixed gate bias voltage is applied the source contact should be located on a virtual ground node and furthermore the drain voltage must at any time stay below the source voltage, both to ensure a good linearity. Small exceptions on previous limits can of course be allowed at the cost of linearity but the elbowroom provided by this resistor architecture is rather limited.

Table 2.5 Summary of the comparative study on resistors in organic electronics technology

Architecture	Metal wire	Pentacene resistor	Linear region transistor
R per square ($\Omega/\square$)	0.8	19	270
R per Area	$20\,\mathrm{k}\Omega/\mathrm{mm}^2$	-	-
Reliability	High	Very low	Moderate
Employability	Low	Low	Moderate

2.7.1.4 Discussion

A resistor that is used in an analog circuit typically has a value that is in the same range or larger than the voltage–current ratio in that circuit. At the moment for organic electronics technology, which suffers from a low mobility compared to standard CMOS, this voltage–current ratio is located in the 1–$100\,\mathrm{M}\Omega$ range. Three different resistor architectures, the metal line resistor, the pentacene resistor and the transistor biased in the linear region have been discussed in Sects. 2.7.1.1 to 2.7.1.3. The metal line resistors are probably the most robust and linear resistors yet they have resistances which are two or three decades below the useful range in organic electronics technology. Consequently, the required area to build the resistors is far too large. Pentacene resistors built with a pure pentacene film have shown a very high resistance per square, which is probably even too high. Moreover, pentacene resistors are expected to depend on the threshold voltage and therefore their resistance is unpredictable and unreliable in organic electronics technology. Finally, transistors biased in the linear regime have been employed as resistors. These give nice measurement results in the desired resistive range. Their drawback is that the linearity is only ensured when $V_{\mathrm{SG}} - V_T$ is larger than V_{SD} and furthermore in practical application the source contact must be connected to a virtual ground node in order to keep V_{SG} fixed. In Table 2.5 a summary of the comparative resistor study is given.

2.7.2 Capacitors

Large rectangular metal plate capacitors can be constructed using the available gold layers in the technology. In Eq. (2.24) the capacitance per area C_A of such a capacitor built with metal layers M_1 and M_2 is given.

$$C_A = 85\,\mathrm{pF/mm}^2 \tag{2.24}$$

Since the permittivity is a fixed material constant, the variations on the capacitance are only determined by the parameter d in an inversely proportional way hence only small relative variations of the capacitance are expected. Furthermore, a parasitic leakage path is experienced in the capacitor. This leakage is experimentally determined and an RC-constant of 15 s has been measured. The capacitor is therefore considered to be a reliable component in the organic electronics technology.

2.7.3 Inductors

Integrated spiral inductors can practically be fabricated in any of the three metal layers in the technology. They even profit from the absence of a conductive substrate so that their quality is relatively high. Nevertheless, due to the low carrier mobility in the transistors and accordingly the high impedances in the circuits very high inductances are needed to employ inductors. The required area to make the adequate inductors is several orders larger than the intended circuits themselves. A practical example that calculates the required inductor for an inductive DC–DC converter is given in Sect. 6.2.2. This example clearly demonstrates that for most of the applications inductors are not much of use.

One exception exists in which a spiral inductor is employed in an organic circuit, i.e., in inductively coupled RFID tags (Myny et al. 2009). There an inductor is driven by radio frequency signals and it drives a rectifier which in turn powers the RFID tag. However, the inductor was not integrated in the system and at the moment no organic circuits with integrated inductors have been published yet.

2.8 Conclusion

Organic thin-film electronics are a promising technology with an outlook to a large niche application field that reaches from flexible displays over flexible lighting to smart sensor systems. The increasing interest for these technologies is a driving force for the development and an exponential increase of the thin-film electronics market is predicted.

The subject of this dissertation is the design of analog circuits in organic electronics technology. It is important for a circuit designer to understand the basic principles of the technology. The focus in this chapter was on the organic thin-film electronics technology. It was the aim to understand the basic principles of thin-film transistors and especially to get a deeper insight into the behavior of transistors and other components in the applied technology in order to optimally use all the available features and to overcome the weak spots that are present.

Organic thin-film transistors perform with a MOS-like transistor behavior that has a saturation regime as well as a subthreshold and a linear regime. However, their physical behavior is totally different from that in Si transistors. They perform in the accumulation regime rather than in inversion and furthermore they are often only practicable for unipolar transistor technologies.

Pentacene is an organic small molecule with a semiconductor behavior. It is used for making p-type transistors and its charge carrier mobility is in the 0.1–$1\,cm^2/Vs$

range. It is one of the best known organic semiconducting materials and most often used for production. It has been already combined with n-type semiconductors for making complementary transistor technologies.

The production of organic electronics technology can be done in several ways. A short overview is given on the most important techniques. Besides thermal vacuum evaporation or vapor-phase deposition also printing techniques exist that push ahead the research on roll-to-roll printing.

The technology applied in this work was extensively discussed. The behavior of both the 3-contact transistor and the 4-contact transistor with its additional backgate were treated. Furthermore, the features of the 4-contact transistor were elucidated through a set of practical application-oriented examples and it was demonstrated that the 4-contact transistor performs better in every situation.

Next a short discussion was held about the non-ideal behavior that is present in the technology. The sensitivity of the behavioral transistor parameters to the production process was investigated and the effects that are present at runtime, such as bias stress and influence by H_2O, O_2 or temperature, were examined. The threshold voltage V_T is the most sensitive behavioral parameter. For analog circuit design V_T is therefore predicted to be the most important parameter to deal with when optimizing the circuit reliability.

Finally, also the realization of passive components in the applied organic electronics technology is examined. Resistors can best be constructed with transistors biased in the linear regime yet their practicability is only moderate. Capacitors are easily constructed with overlapping metal layers.

References

Bai Y.-W., Chen C.-Y (2007) Using serial resistors to reduce the power consumption of resistive touch panels. In: IEEE International Symposium on Consumer Electronics ISCE 2007, pp. 1–6.

Bode D, Myny K, Verreet B, van der Putten B, Bakalov P, Steudel S, Smout S, Vicca P, Genoe J, Heremans P (2010) Organic complementary oscillators with stage-delays below 1 μs. Appl Phys Lett 96(13):133307

Bonse M, Thomasson DB, Klauk H, Gundlach DJ, Jackson TN (1998) Integrated a-Si:H/pentacene inorganic/organic complementary circuits. In: International technical digest in electron devices meeting IEDM '98, pp. 249–252.

Caboni A, Orgiu E, Barbaro M, Bonfiglio A (2009) Flexible organic Thin-Film transistors for pH monitoring. IEEE Sens J 9(12):1963–1970

Cantatore E, Geuns TCT, Gelinck GH, van Veenendaal E, Gruijthuijsen AFA, Schrijnemakers L, Drews S, de Leeuw DM (2007) A 13.56-MHz RFID system based on organic transponders. IEEE J Solid-State Circuits, 42(1):84–92.

Chen H-Y, Patil N, Lin A, Wei L, Beasley C, Zhang J, Chen X, Wei H, Liyanage LS, Shulaker MM, Mitra S, Philip Wong H-S (2011) Carbon electronics from material synthesis to circuit demonstration. In: International symposium on VLSI technology, systems and applications (VLSI-TSA) 2011, pp. 1–2.

Chua L-L, Zaumseil J, Chang J-F, Ou ECW, Ho PKH, Sirringhaus H, Friend RH (2005) General observation of n-type field-effect behaviour in organic semiconductors. Nature 434(7030):194–199

Chuo Y, Omrane B, Landrock C, Aristizabal J, Hohertz D, Niraula B, Grayli SV, Kaminska B (2011) Powering the future: Integrated, thin, flexible organic solar cells with polymer energy storage. In. IEEE Design and Test of Computers, vol PP(99), p 1.

Comiskey B, Albert JD, Yoshizawa H, Jacobson J (2005) An electrophoretic ink for all-printed reflective electronic displays. Nature 394(6690):253–255

Darlinski G, Böttger U, Waser R, Klauk H, Halik M, Zschieschang U, Schmid G, Dehm C (2005) Mechanical force sensors using organic thin-film transistors. J Appl Phys 97(9):093708

Debucquoy M, Verlaak S, Steudel S, Myny K, Genoe J, Heremans P (2007) Correlation between bias stress instability and phototransistor operation of pentacene thin-film transistors. Appl Phys Lett 91(10):103508-1-103508-3.

Debucquoy M, Rockele M, Genoe J, Gelinck GH, Heremans P (2009) Charge trapping in organic transistor memories: On the role of electrons and holes. Org Electron 10(7):1252–1258

de Leeuw DM, Simenon MMJ, Brown AR, Einerhand REF (1997) Stability of n-type doped conducting polymers and consequences for polymeric microelectronic devices. Synthetic Metals 87:53–59

De Vusser S, Steudel S, Myny K, Genoe J, Heremans P (2006) A 2 V organic complementary inverter. In: IEEE international solid-state circuits conference ISSCC 2006, Digest of Technical Papers. pp 1082–1091.

Diallo K, Erouel M, Tardy J, Andre E, Garden J-L (2007) Stability of pentacene top gated thin film transistors. Appl Phys Lett 91(18):183508-183508-3.

Dickson JF (1976) On-chip high-voltage generation in MNOS integrated circuits using an improved voltage multiplier technique. IEEE J Solid-State Circuits 11(3):374–378

Dimitrakopoulos CD, Malenfant PRL (2002) Organic thin film transistors for large area electronics. Adv Mater 14(2):99–117

Feng Y, Lee K, Farhat H, Kong J (2009) Current on/off ratio enhancement of field effect transistors with bundled carbon nanotubes. J Appl Phys 106(10):104505

Fujisaki Y, Nakajima Y, Kumaki D, Yamamoto T, Tokito S, Kono T, Nishida JI, Yamashita Y (2010) Air-stable n-type organic thin-film transistor array and high gain complementary inverter on flexible substrate. Appl Phys Lett 97(13):133303

Gay N, Fischer W-J (2007) OFET-based analog circuits for microsystems and RFID-sensor transponders. In: 6th international conference on polymers and adhesives in microelectronics and photonics, polytronic 2007, pp 143–148.

Gelinck GH, Huitema HEA, van Veenendaal E, Cantatore E, Schrijnemakers L, van der Putten JBPH, Geuns TCT, Beenhakkers M, Giesbers JB, Huisman B-H, Meijer EJ, Benito EM, Touwslager FJ, Marsman AW, van Rens BJE, de Leeuw DM (2004) Flexible active-matrix displays and shift registers based on solution-processed organic transistors. Nat Mater 3(2):106–110

Gelinck GH, van Veenendaal E, Coehoorn R (2005) Dual-gate organic thin-film transistors. Appl Phys Lett 87(7):073508

Gelinck G, Heremans P, Nomoto K, Anthopoulos TD (2010) Organic transistors in optical displays and microelectronic applications. Adv Mater (Deerfield Beach, Fla.), 22(34):3778–3798.

Genoe J, Steudel S, De Vusser S, Verlaak S, Janssen D, Heremans P (2004) Bias stress in pentacene transistors measured by four probe transistor structures. In: Proceeding of the 34th European solid-state device research conference ESSDERC 2004, pp 413–416.

Gross L, Mohn F, Moll N, Liljeroth P, Meyer G (2009) The chemical structure of a molecule resolved by atomic force microscopy. Science 325(5944):1110–1114

Guo D, Miyadera T, Ikeda S, Shimada T, Saiki K (2007) Analysis of charge transport in a polycrystalline pentacene thin film transistor by temperature and gate bias dependent mobility and conductance. J Appl Phys 102(2):023706

Halik M, Klauk H, Zschieschang U, Kriem T, Schmid G, Radlik W, Wussow K (2002) Fully patterned all-organic thin film transistors. Appl Phys Lett 81:289

He DD, Nausieda IA, Ryu KK, Akinwande AI, Bulovic V, Sodini CG (2010) An integrated organic circuit array for flexible large-area temperature sensing. In: IEEE international solid-state circuits conference digest of technical papers (ISSCC) 2010, pp 142–143.

Herwig PT, Mllen K (1999) A soluble pentacene precursor: synthesis, solid-state conversion into pentacene and application in a field-effect transistor. Adv Mater 11(6):480–483

Heremans P, Myny K, Marien H, van Veenendaal E, Steudel S, Steyaert M (2010) Gelinck G (2010) Application of organic thin-film transistors for circuits on flexible foils. Proceedings of international display workshop, Fukuoka, In

Heremans P, Dehaene W, Steyaert M, Myny K, Marien H, Genoe J, Gelinck G, van Veenendaal E (2011) Circuit design in organic semiconductor technologies. In: Proceedings of the ESSCIRC (ESSCIRC) 2011, pp 5–12.

Horowitz G, Fichou D, Peng X, Xu Z, Garnier F (1989) A field-effect transistor based on conjugated alpha-sexithienyl. Solid State Commun. 72(4):381–384

Huitema HEA et al. (2008) Rollable displays: the start of a new mobile device generation. In: 7th annual USDC flexible electronics and display conference, Phoenix, January 2008.

IDTechEx (2011) Printed, organic and flexible electronics forecasts, players and opportunities 2011–2021, http://www.idtechex.com/, 2011

Jang S-J, Ahn J-H (2010) Flexible thin film transistor using printed single-walled carbon nanotubes. In: 3rd international nanoelectronics conference (INEC) 2010, pp 720–721.

Jeong SW, Jeong JW, Chang S, Kang SY, Cho KI, Ju BK (2010) The vertically stacked organic sensor-transistor on a flexible substrate. Appl Phys Lett 97(25):253309

Jurchescu OD, Baas J, Palstra TTM (2005) Electronic transport properties of pentacene single crystals upon exposure to air. Appl Phys Lett 87(5):052102–052102-3.

Kagan CR, Afzali A, Graham TO (2005) Operational and environmental stability of pentacene thin-film transistors. Applied Physics Letters 86(19):193505

Kane MG, Campi J, Hammond MS, Cuomo FP, Greening B, Sheraw CD, Nichols JA, Gundlach DJ, Huang JR, Kuo CC, Jia L, Klauk H, Jackson TN (2000) Analog and digital circuits using organic thin-film transistors on polyester substrates. IEEE Electron Device Lett 21(11):534–536

Kawaguchi H, Someya T, Sekitani T, Sakurai T (2005) Cut-and-paste customization of organic FET integrated circuit and its application to electronic artificial skin. IEEE J Solid-State Circuits 40(1):177–185

Kim H-K, Lee S-G, Han J-E, Kim T-R, Hwang S-U, Ahn SD, You I-K, Cho K-I, Song T-K, Yun K-S (2009) Transparent and flexible tactile sensor for multi touch screen application with force sensing. In: International solid-state sensors, actuators and microsystems conference TRANSDUCERS 2009, pp 1146–1149.

Klauk H (2006) Organic electronics, materials, manufacturing and applications. Wiley-VCH Verlag GmbH and Co KGaA, Weinheim

Lapinski M, Domaradzki J, Prociow EL, Sieradzka K, Gornicka B (2009) Electrical and optical characterization of ITO thin films. In: International students and young scientists workshop "Photonics and Microsystems" 2009, pp 52–55.

Li L, Debucquoy M, Genoe J, Heremans P (2010) A compact model for polycrystalline pentacene thin-film transistor. J Appl Phys 107:024519

Li L, Marien H, Genoe J, Steyaert M, Heremans P (2010) Compact model for organic Thin-Film transistor. IEEE Electron Device Lett 31(3):210–212

Liu B, Xie G-Z, Du X-S, Li X, Sun P (2009) Pentacene based organic thin-film transistor as gas sensor. In: International conference on apperceiving computing and intelligence analysis. ICACIA 2009:1–4

Maddalena F, Spijkman M, Brondijk JJ, Fonteijn P, Brouwer F, Hummelen JC, de Leeuw DM, Blom PWM, de Boer B (2008) Device characteristics of polymer dual-gate field-effect transistors. Org Electron 9(5):839–846

Manunza I, Sulis A, Bonfiglio A (2006) Pressure sensing by flexible, organic, field effect transistors. Appl Phys Lett 89(143502):1–3

Marien H, Steyaert M, van Aerle N, Heremans P (2009) A mixed-signal organic 1 kHz comparator with low VT sensitivity on flexible plastic substrate. In: Proceedings of ESSCIRC '09, pp 120–123.

Marien H, Steyaert M, Steudel S, Vicca P, Smout S, Gelinck G, Heremans P (2010) An organic integrated capacitive DC-DC up-converter. In: Proceedings of the ESSCIRC 2010, pp 510–513.

Marien H, Steyaert M, van Aerle N, Heremans P (2010) An analog organic first-order CT $\Delta\Sigma$ ADC on a flexible plastic substrate with 26.5 dB precision. In: IEEE international solid-state circuits conference digest of technical papers (ISSCC) 2010, pp 136–137.

Marien H, Steyaert M, van Veenendaal E, Heremans P (2010) Analog techniques for reliable organic circuit design on foil applied to an 18 dB single-stage differential amplifier. Org Electron 11(8):1357–1362

Marien H, Steyaert M, van Veenendaal E, Heremans P (2011) DC-DC converter assisted two-stage amplifier in organic thin-film transistor technology on foil. In: Proceedings of the ESSCIRC 2011, pp 411–414.

Marien H, Steyaert M, van Veenendaal E, Heremans P (2011) Organic dual DC-DC upconverter on foil for improved circuit reliability. Electron. Lett. 47(4):278–280

Marien H, Steyaert MSJ, van Veenendaal E, Heremans P (2011) A fully integrated $\Delta\Sigma$ ADC in organic thin-film transistor technology on flexible plastic foil. IEEE J Solid-State Circuits 46(1):276–284

Marien H, Steyaert M, van Veenendaal E (2011) Heremans P (2011) ADC design in organic Thin-Film electronics technology on plastic foil. International workshop on ADC (IWADC), Orvieto, June, In.

Marien H, Steyaert M, van Veenendaal E, Heremans P (2012) 1D and 2D analog 1.5 kHz air-stable organic capacitive touch sensors on plastic foil. In: IEEE international solid-state circuits conference digest of technical papers (ISSCC) 2012.

Daily Mobile (2012) Video: Samsung flexible amoted display (ces 2011), http://forum.dailymobile. se. Accessed 08 March 2012

Molesa SE, Volkman SK, Redinger DR, Vornbrock AdF, Subramanian V (2004) A high-performance all-inkjetted organic transistor technology. In: IEEE international electron devices meeting IEDM Technical Digest 2004, pp 1072–1074.

Mori T, Kikuzawa Y, Noda K (2009) Improving the sensitivity and selectivity of alcohol sensors based on organic thin-film transistors by using chemically-modified dielectric interfaces. In: IEEE sensors 2009, pp 1951–1954.

Myny K, Steudel S, Vicca P, Beenhakkers MJ, van Aerle NAJM, Gelinck GH, Genoe J, Dehaene W, Heremans P (2009) Plastic circuits and tags for 13.56 MHz radio-frequency communication. Solid-State Electron. 53(12):1220–1226

Myny K, Beenhakkers MJ, van Aerle NAJM, Gelinck GH, Genoe J, Dehaene W, Heremans P (2011) Unipolar organic transistor circuits made robust by Dual-Gate technology. IEEE J Solid-State Circuits 46(5):1223–1230

Myny K, Rockel M, Chasin A, Pham D-V, Steiger J, Botnaras S, Weber D, Herold B, Ficker J, van der Putten B, Gelinck GH, Genoe J, Dehaene W (2012) Heremans P (2012) Bidirectional communication in an hf hybrid organic/solution-processed metal-oxide rfid tag. IEEE international solid-state circuits conference digest of technical papers (ISSCC), In.

Na JH, Kitamura M, Arakawa Y (2008) Organic/inorganic hybrid complementary circuits based on pentacene and amorphous indium gallium zinc oxide transistors. Appl. Phys. Lett. 93(21):213505

Na JH, Kitamura M, Arakawa Y (2009) Low-voltage-operating organic complementary circuits based on pentacene and C60 transistors. Thin Solid Films 517(6):2079–2082

Ng TN, Chabinyc ML, Street RA, Salleo A (2007) Bias stress effects in organic thin film transistors. In: Proceedings of the 45th annual IEEE international reliability physics, symposium 2007, pp 243–247.

Morooka M, Arai T, Morosawa N, Ohshima Y, Sasaoka Y (2011) A novel self-aligned top-gate oxide TFT for AM-OLED displays. In: Proceedings of the SID 11 Digest 2011, pp 479–482.

Noda M, Kobayashi N, Katsuhara M, Yumoto A, Ushikura S (2010) A rollable AM-OLED display driven by OTFTs. In: Proceedings of the SID 2010, vol 41, pp 710–713.

Novoselov KS, Geim AK, Morozov SV, Jiang D, Zhang Y, Dubonos SV, Grigorieva IV, Firsov AA (2004) Electric field effect in atomically thin carbon films. Science 306(5696):666–669

Oberhoff D, Pernstich KP, Gundlach DJ, Batlogg B (2007) Arbitrary density of states in an organic Thin-Film Field-Effect transistor model and application to pentacene devices. IEEE Trans Electron Devices 54(1):17–25

UK Telematics Online, "What is rfid?)", http://www.uktelematicsonline.co.uk, Accessed 08 March 2012

University of Tokyo, http://www.ntech.t.u-tokyo.ac.jp, Accessed 08 March 2012

Pulse on Tech, "The readius. worlds first pocket ereader", http://www.pulseontech.com, Accessed 08 March 2012

Pannemann Ch, Diekmann T, Hilleringmann U (2004) On the degradation of organic field-effect transistors. In: Proceedings of the 16th international conference on microelectronics ICM 2004, pp 76–79.

Pannemann Ch, Diekmann T, Hilleringmann U, Schurmann U, Scharnberg M, Zaporojtchenko V, Adelung R, Faupel F (2005) Encapsulating the active layer of organic Thin-Film transistors. In: 5th international conference on polymers and adhesives in microelectronics and photonics, polytronic 2005, pp 63–66.

Pelgrom MJM, Duinmaijer ACJ, Welbers APG (1989) Matching properties of MOS transistors. IEEE J Solid-State Circuits 24(5):1433–1439

Pernstich KP, Oberhoff D, Goldmann C, Batlogg B (2006) Modeling the water related trap state created in pentacene transistors. Appl Phys Lett 89(21):213509–213509-3.

Rolin C, Steudel S, Myny K, Cheyns D, Verlaak S, Genoe J, Heremans P (2006) Pentacene devices and logic gates fabricated by organic vapor phase deposition. Appl Phys Lett 89(20):203502

Rolin C, Vasseur K, Schols S, Jouk M, Duhoux G, Muller R, Genoe J, Heremans P (2008) High mobility electron-conducting thin-film transistors by organic vapor phase deposition. Appl Phys Lett 93(3):033305

Sigma-Aldrich, "Safety data sheet: Pentacene", http://www.sigma-aldrich.com/, 2010

Sancho-Garca JC, Horowitz G, Bredas JL, Cornil J (2003) Effect of an external electric field on the charge transport parameters in organic molecular semiconductors. J Chem Phys 119(23):12563

Sansen MCW (2008) Analog Design Essentials. Springer, Berlin

Raf Schoofs, (2007) Design of High-Speed Continuous-Time Delta-Sigma A/D Converters for Broadband Communication, Ph.D. Thesis, ESAT-MICAS, K.U.Leuven, Belgium.

Schwierz F (2010) Graphene transistors a new contender for future electronics. In: 10th IEEE international conference on solid-state and integrated circuit technology (ICSICT) 2010, pp 1202–1205.

Shafiee A, Salleh MM, Yahaya M (2008) Fabrication of organic solar cells based on a blend of donor-acceptor molecules by inkjet printing technique. In: IEEE international conference on semiconductor electronics. ICSE 2008:319–322

Spijkman M-J, Myny K, Smits ECP, Heremans P, Blom PWM, de Leeuw DM (2011) Dual-gate Thin-Film transistors integrated circuits and sensors. Adv Mater 23(29):3231–3242

Street RA, Chabinyc ML, Endicott F, Ong B (2006) Extended time bias stress effects in polymer transistors. J Appl Phys 100(11):114518–114518-10.

Subramanian V, Lee JB, Liu VH, Molesa S (2006) Printed electronic nose vapor sensors for consumer product monitoring. In: IEEE international solid-state circuits conference ISSCC 2006, Digest of Technical Papers. pp 1052–1059.

Van Breussegem T, Steyaert M (2010) A fully integrated 74% efficiency 3.6 V to 1.5 V 150 mW capacitive point-of-load DC/DC-converter. In: Proceedings of the ESSCIRC 2010, pp. 434–437.

Van Oost J (2011) Universiteit hasselt vindt printbaar zonnepaneel uit", http://www.zdnet.be, 2011

Verlaak S, Steudel S, Heremans P, Janssen D, Deleuze MS (2003) Nucleation of organic semiconductors on inert substrates. Phys Rev B 68(19):195409

Vollmer A, Weiss H, Rentenberger S, Salzmann I, Rabe JP, Koch N (2006) The interaction of oxygen and ozone with pentacene. Surf Sci 600(18):4004–4007

Wang L, Li D, Hu Y, Jiang C (2011) Realization of uniform large-area pentacene thin film transistor arrays by roller vacuum thermal evaporation.

Wang T-M, Ker M-D (2011) Design and implementation of capacitive sensor readout circuit on glass substrate for touch panel applications. In: International symposium on VLSI design automation and test (VLSI-DAT) 2011, pp 1–4.

Wens M, Cornelissens K, Steyaert M (2007) A fully-integrated 0.18 m CMOS DC-DC step-up converter, using a bondwire spiral inductor. In: 33rd European solid state circuits conference ESSCIRC 2007, pp 268–271.

Wens M, Steyaert MSJ (2011) A fully integrated CMOS 800-mW Four-Phase semiconstant ON/OFF-Time Step-Down converter. IEEE Trans Power Electron 26(2):326–333

Wikipedia (2011) Wikipedia: The free encyclopedia. http://www.wikipedia.org/, 2011

Wilson TV (2007) How the iPhone works. http://electronics.howstuffworks.com/iphone.htm, 2007

Wilke B, Gobel H (2008) Investigations on the temperature dependence of the charge carrier mobility of pentacene field-effect transistors. In: 2nd IEEE international interdisciplinary conference on portable information devices, 2008 and 7th IEEE conference on polymers and adhesives in microelectronics and photonics. PORTABLE-POLYTRONIC 2008:1–5

Lo HW, Tai Y-C (2007) Characterization of parylene as a water barrier via buried-in pentacene moisture sensors for soaking tests. In: 2nd IEEE international conference on Nano/Micro engineered and molecular systems NEMS '07, pp 872–875.

Wrachien N, Cester A, Bellaio N, Pinato A, Meneghini M, Tazzoli A, Meneghesso G, Myny K, Smout S, Genoe J (2010) Light, bias, and temperature effects on organic TFTs. In: IEEE international reliability physics symposium (IRPS) 2010, pp 334–341.

Xiong W, Guo Y, Zschieschang U, Klauk H, Murmann B (2010) A 3-V, 6-Bit C-2C Digital-to-Analog converter using complementary organic Thin-Film transistors on glass. IEEE J Solid-State Circuits 45(7):1380–1388

Xiong W, Zschieschang U, Klauk H, Murmann B (2010) A 3V 6b successive-approximation ADC using complementary organic thin-film transistors on glass. In: IEEE international solid-state circuits conference digest of technical papers (ISSCC) 2010, pp 134–135.

Yokota T, Sekitani T, Tokuhara T, Zschieschang U, Klauk H, Huang T-C, Takamiya M, Sakurai T, Someya T (2011) Sheet-type organic active matrix amplifier system using Vth-Tunable. Pseudo-CMOS circuits with floating-gate structure.

Zan HW, Tsai WW, Lo YR, Wu YM, Yang YS (2011) Pentacene-based organic thin film transistors for ammonia sensing. IEEE Sensors J PP(99):1.

Chapter 3
Amplifier Design

The implemention of smart sensor systems requires besides a digital RFID circuit, also sensors, amplifiers and analog-to-digital interfacing circuits. Those are analog building blocks that have to meet certain specifications. The differential amplifier is probably the most important basic analog building block that is employed in almost every analog circuit. It is a building block with low complexity but its implementation depends on the technological environment in which the amplifier is employed. This environment, presented in Chap. 2, is atypical compared to the existing standard *Si* technology and several design challenges show up. Therefore the design of a differential amplifier in organic electronics technology is not straightforward and requires special attention. In organic technology, which is bounded by certain limits on the performance, it is essential to exploit all the advantages and to overcome all the challenges that are present. The unipolar nature of the applied technology, for instance, has an impact on the specifications of the amplifier and hampers the application of complex circuit techniques. Other features of the technology, e.g. the backgate, are beneficial in certain situations.

At the moment only a few papers have discussed basic analog organic circuits (Kane et al. 2000; Gay and Fischer 2007). Kane et al. (2000) presents a differential ring oscillator. In Gay and Fischer (2007) a model is presented for organic transistors on a glass substrate and this is applied to a few basic circuits.

In Fig. 1.5 the architecture of an organic smart sensor system is proposed and the amplifier block is the subject of this chapter. In this chapter step-by-step the design methodology of organic differential amplifiers is discussed. First the application field is highlighted in Sect. 3.1. Then a topological study of the organic amplifier is given in Sect. 3.2 where special attention goes to the implementation of p-type load transistors. Consequently in Sect. 3.3 some designs and measurement results are presented: a single-stage differential amplifier, a three-stage operational amplifier, an improved amplifier and a comparator. Finally this chapter is concluded in Sect. 3.4.

H. Marien et al., *Analog Organic Electronics*, Analog Circuits and Signal Processing, 59
DOI: 10.1007/978-1-4614-3421-4_3, © Springer Science+Business Media New York 2013

3.1 Application Field

The application field of integrated amplifiers is endless. They are used as a building block in almost every larger analog or mixed-signal circuit. The same goes for amplifiers in organic electronics. Although most of the applications in the domain of organic electronics are digital, also analog and mixed-signal applications exist. In the organic smart sensor systems, for instance, a post-amplifier is required that amplifies the output signal of the sensor. Furthermore, analog-to-digital conversion is required that converts the measured signal to digital data and that typically employs differential amplifiers in feedback loops.

The deduction of design specifications for the amplifier is rather easy.

In the first place the amplifier should reach as much gain as possible. Since the intrinsic gain of a transistor is relatively low and only p-type transistors are available it is a challenge for the designer to reach a high gain. Therefore the gain is the specification with the highest priority.

In the second place comes reliability. This means that the amplifier must be designed in a defensive way that deals with the expected technological and behavioral variations discussed in Sect. 2.5. This defensive strategy interferes with the high priority of the gain specification, hence both specifications are in a trade-off that must be well balanced to reach a good and reliable behavior.

Finally, in the third place, it is important for the amplifier to have a certain signal bandwidth. In this study no specific attention is paid to the bandwidth of the amplifiers. This would bring about another trade-off with the gain. From the technological point of view a lot of research is still being done in order to increase the carrier mobility in organic semiconductors. Therefore the bandwidth of circuits will automatically increase in time, simultaneously with the progress made with the carrier mobility at the technology level.

3.2 Topology

The technological environment in which the amplifiers are designed is a limiting factor for their quality. An intelligent topological choice and well-chosen circuit techniques, however, overcome certain sensitivities directly at the circuit level. In this section the amplifier architectures are investigated and compared for their ability to deal with the unipolar character of the technology and with the variability of the threshold voltage V_T that is typically present in organic electronics technology. The ΔV_T change between two processing runs, separated by a long time, i.e. a few months, is one of the dominating challenges for amplifier design. An objective figure of merit to compare the amplifier topologies for their sensitivity to such a ΔV_T, the threshold voltage suppression ratio (VTSR), is therefore presented.

In this section first a single-stage single-ended amplifier and a single-stage differential amplifier are discussed. A closer look of the load transistors of the unipolar

Fig. 3.1 Schematic view
and small-signal equivalent
scheme of a single-ended
amplifier. r_L is the output
resistance of a non-ideal
current source I_L

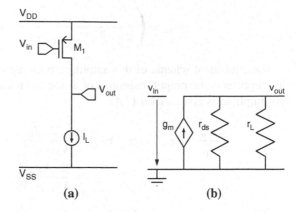

(a)   (b)

amplifier for which p-type solutions are provided, follows. After a behavioral comparison also simulation results are presented that compare differential amplifiers, all built with a different load topology. Finally the applicability of a set of standard circuit techniques, such as cascoding and gain boosting, is discussed.

3.2.1 Single-Ended Amplifier

The p-type single-ended amplifier is presented in Fig. 3.1. This amplifier is built up with one p-type transistor M_1 and a current source. The DC voltage of the output node is determined by the component equations of both the transistor M_1 and the current source. The component equation of M_1 biased in saturation is given in Eq. (3.1):

$$I_{SD} = K'_p \cdot \frac{W}{L}(V_{SG} - V_T)^2 \left(1 + \frac{V_{SD}}{V_E \cdot L}\right) \tag{3.1}$$

where K'_p is a technology-defined constant and V_E is the Early voltage of the transistor. The output voltage of the amplifier is obtained by rewriting this equation into Eq. (3.2):

$$V_{out} = V_{DD} - V_{SD} = V_{DD} - \left(\frac{I_L}{K'_p \cdot \frac{W}{L}(V_{SG} - V_T)^2} - 1\right)V_E \cdot L \tag{3.2}$$

An acceptable value for the DC level of V_{out} is $V_{DD}/2$. Then the equation is slightly simplified to Eq. (3.3):

$$V_{out} = \left(\frac{I_L}{K'_p \cdot \frac{W}{L}(V_{SG} - V_T)^2} - 1 \right) V_E \cdot L \tag{3.3}$$

The equivalent scheme of this amplifier, presented in Fig. 3.1 uses a resistor r_L that represents the output resistance of a non-ideal current source I_L. The gain of this amplifier is given by Eq. (3.4):

$$\frac{v_{out}}{v_{in}} = g_{m,1} \cdot (r_L \parallel r_{0,1}) = \frac{2 \cdot I_C}{V_{SG} - V_T} \cdot \left(r_L \parallel \frac{V_E \cdot L}{I_C} \right)$$
$$\approx \frac{2 \cdot V_E \cdot L}{V_{SG} - V_T} \tag{3.4}$$

The gain can be increased by biasing the transistor with a small gate overdrive $V_{SG} - V_T$ and, on the supposition that r_L is high, by increasing L of the transistor while keeping the W/L ratio constant.

As discussed in Sect. 2.5 the V_T of the transistors in organic electronics technology is a parameter that is sensitive to a set of process and environmental variables. Therefore it is important to investigate and compare the sensitivity of the amplifier topologies to variations of this parameter. In both Eqs. (3.2) and (3.4) the factor $(V_{SG} - V_T)$ is in the denominator. Moreover this factor is small since the transistor M_1 is typically biased with V_{SG} close to V_T for high gain. Therefore both the DC output voltage level and the gain are very sensitive to a ΔV_T change.

A figure of merit that is presented in this work to compare the sensitivity of an amplifier circuit to ΔV_T in the input transistors is the threshold voltage suppression ratio (VTSR) derived in Eq. (3.5). It represents the ratio between the gain of the amplifier and the common-mode gain of a ΔV_T excitation in the input transistors.

$$VTSR = \left| \frac{\frac{\partial V_{out}}{\partial V_{in}}}{\frac{\partial V_{out}}{\partial V_{T,in}}} \right| = \left| \frac{A}{A_{V_T}} \right| \tag{3.5}$$

$$VTSR_{se} = \left| \frac{\frac{\partial V_{out}}{\partial V_{in}}}{\frac{\partial V_{out}}{\partial V_{in}}} \right| = 1 = 0 dB \tag{3.6}$$

where A and A_{V_T} are respectively the gain of the amplifier and the V_T gain of the amplifier. In this formula only the ΔV_T of the input transistors is included. This is clarified in Sect. 3.2.3 where the implementation of the load transistors is profoundly discussed. Since the current through a transistor in saturation is always a function of the gate overdrive $V_{SG} - V_T$, every ΔV_T in M_1 of the single-ended amplifier has an identical effect to a $-\Delta V_{in}$ of the amplifier. Therefore the VTSR$_{se}$ for a single-ended amplifier is 1 or 0 dB, hence the higher the gain, the higher the V_T sensitivity of this amplifier. In this topology there is no degree of freedom that separates the amplifier behavior from the ΔV_T sensitivity, whereas there is such an opportunity in the differential amplifier that is discussed in Sect. 3.2.2.

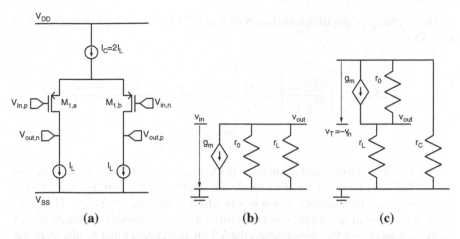

Fig. 3.2 Schematic view of a differential amplifier and equivalent schemes for differential mode and common mode. r_L and r_C are the output resistances of I_L and I_C

3.2.2 Differential Amplifier

Whereas in a single-ended amplifier the gain and the ΔV_T sensitivity are strictly related parameters, these two behaviors are separated in a differential amplifier. The gain of the amplifier is now determined in the differential-mode behavior whereas the effect of a ΔV_T shift shows up in the common-mode behavior. In Fig. 3.2 the schematic view of a differential amplifier is presented together with its small-signal equivalent scheme for both the differential-mode and the common-mode behavior. The small-signal resistors r_L and r_C represent the output resistances of the non-ideal current sources I_C and I_L.

$$A_d = \frac{v_{out}}{v_{in}} = g_m \cdot (r_0 \parallel r_L) \approx g_m \cdot r_0 \qquad (3.7)$$

$$A_{V_T} = \frac{v_{out}}{v_{T,in}} = \frac{g_m \cdot r_0 \cdot r_L}{r_0 + r_L + r_C + g_m \cdot r_0 \cdot r_C}$$

$$\approx \frac{-r_L}{r_C} \qquad (3.8)$$

The differential gain of the amplifier is given by Eq. (3.7), which is identical to the gain of the single-ended amplifier. The effect of a ΔV_T change, however, differs from the gain and is derived in Eq. (3.8). This formula is rather complex but it is clear that for the case of a high r_C, the influence of a ΔV_T in M_1 goes to 0. On the other hand when r_C is very low, the behavior of the differential amplifier reverts to that of the single-ended amplifier.

The VTSR$_{dp}$ of the differential amplifier in Eq. (3.9) is derived from Eqs. (3.7) and (3.8):

$$VTSR_{dp} = \left| \frac{\frac{\partial V_{out}}{\partial V_{in}} |_{dm}}{\frac{\partial V_{out}}{\partial V_{in}} |_{cm}} \right| = \frac{A_d}{A_{V_T}} = \frac{r_0 + r_L + r_C + g_m \cdot r_0 \cdot r_C}{r_L}$$

$$\approx g_m \cdot r_0 \cdot \frac{r_C}{r_L} \qquad (3.9)$$

The VTSR$_{dp}$ of a differential amplifier is identical to the common mode rejection ratio (CMRR) of the amplifier. This speaks for itself since the current equation of a transistor biased in saturation region is a function of the gate overdrive. Therefore every ΔV_T, present in both the input transistors M_2, corresponds mathematically to a common-mode $-\Delta V_{in}$. Nevertheless the VTSR is a concept that results from the ΔV_T issues and that can be further extrapolated towards the ΔV_T of other transistors in the amplifier circuit.

3.2.3 Load Transistor

In the previous sections current sources I_L were applied as a load for both the single-ended and the differential amplifier. In a complementary technology a p-type single-stage differential amplifier, shown in Fig. 3.3a, applies a p-type input transistor pair M_2, a p-type current transistor M_1 and n-type load transistors M_3. The advantage of these n-type load transistors is that they are connected to the output node through their drain contact. Therefore the small-signal output resistance $r_{sd,3}$ of the n-type is high which is beneficial for the gain of the amplifier. Moreover the behavior of the n-type load transistors is mostly determined by their gate-source voltage V_{GS}, that is determined by a bias voltage, and is relatively insensitive to signals present at their drain contacts, i.e. the output nodes. This bias voltage for its part is the ideal node for applying common-mode feedback in this amplifier.

At the moment in the applied organic electronics technologies n-type transistors are not available and all the advantages mentioned in the previous paragraph are cancelled, hence p-type solutions are required for the implementation of the load transistors. In this section a zero-V_{GS} load, a diode-connected load, a hybrid load and a load with bootstrapped gain-enhancement are discussed and their benefits are compared.

3.2.3.1 Zero-V_{GS} Load

There are two obvious connection schemes for a p-type transistor employed as a load transistor, i.e.

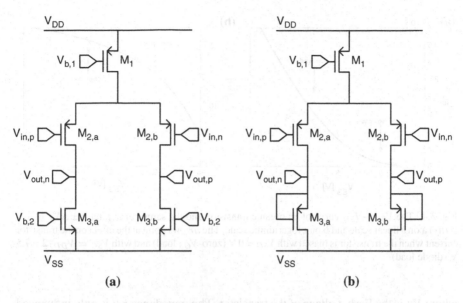

Fig. 3.3 **a** A CMOS differential amplifier with n-type load transistors. **b** A unipolar differential amplifier with p-type zero-V_{GS} load transistors

- the zero-V_{GS} load, where the gate is connected to the source,
- the diode connected load where the gate is connected to the drain.

The schematic view of a p-type differential amplifier with p-type zero-V_{GS} load is presented in Fig. 3.3b. The zero-V_{GS} load is constructed with a p-type transistor of which the gate and source contacts are connected. The drain contact of this transistor is connected to the ground voltage V_{SS} while the gate and source contacts are connected to the output voltage of the differential amplifier. This load topology is commonly used in single-ended digital organic circuits where it enables a low-current solution for logic gates (Myny et al. 2011). In order to estimate the applicability of this load topology in analog circuits it is investigated for two critical aspects, i.e. the small-signal impedance that is experienced at the output node, which determines the gain of the amplifier, and the reliability of the load, i.e. how sensitive this load topology is to the technological and environmental challenges that are present in the topology and are profoundly discussed in Chap. 2.

In terms of output impedance this zero-V_{GS} load topology performs very well. The output resistance that is experienced is r_{sd} of the transistor. This high output resistance enables an amplifier with zero-V_{GS} load to reach a gain, given by Eq. (3.7). The product of r_{sd} and I_{sd} and the gain A_{zvl} of the zero-V_{GS} amplifier are determined by Eqs. (3.10) and (3.11):

$$r_{sd} = \frac{V_E \cdot L}{I_{sd}} \tag{3.10}$$

$$A_{zvl} = g_{m,2} \cdot (r_{sd,2} \parallel r_{sd,3}) \tag{3.11}$$

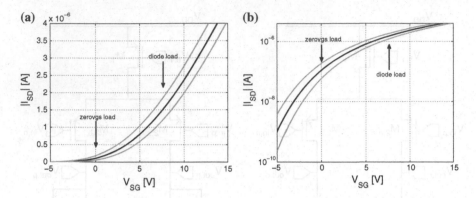

Fig. 3.4 The $V_{SG} - I_{SD}$ curve of an organic transistor and the same curve after a $\pm\Delta V_T$ of 1 V (*gray*) **a** on a linear scale and **b** on a logarithmic scale. The *arrows* point at the effect on the transistor current when the transistor is biased with $V_{SG} = 0$ V (zero-V_{GS} load) and with $V_{SG} = V_{DD}/2 = 7.5$ V (diode load)

where V_E is the Early voltage of the transistor. The impedance r_{sd} is only influenced by $V_{SG} - V_T$ through the current I_{sd}. Consequently, since gain is independent of the current, the gain of the amplifier is stable and high. If so desired the output resistance of the transistor can be further increased by increasing L and keeping the W/L ratio constant.

Although the zero-V_{GS} load topology scores well for high gain, its behavior is not very reliable. In this topology the gate contact and the source contact are connected, so V_{SG} is. The generic $V_{SG} - I_{SD}$ curve of an organic transistor is presented in Fig. 3.4, both in the linear and the logarithmic scale. The gray lines on the curve represent the initial curve that has undergone a shift of the threshold voltage $\pm\Delta V_T$. The effect of $\pm\Delta V_T$ on the current through the zero-V_{GS} transistor is large and on the logarithmic plot it is visible that a $\pm\Delta V_T$ of 1 V can even influence the current by a factor of almost 2. Digital gates are mostly affected by this change in terms of speed, but the proper functionality will remain as long as the gain of the stage remains higher than 1. For analog circuits, such as the differential amplifier, such a V_T change is very adverse for the DC bias levels and a fast degradation of the functionality would occur.

Next to the important case of the ΔV_T change, also the mismatch between the load transistor currents plays an important role in the differential amplifier, i.e. mismatch results in output offset and disturbs the differential output signal of the amplifier. The mismatch in this load topology is therefore a second reason why this zero-V_{GS} load performs poorly. Simulation results of a differential amplifier with this load type employed are given and compared to the simulation results of the other topologies in Sect. 3.2.3.5.

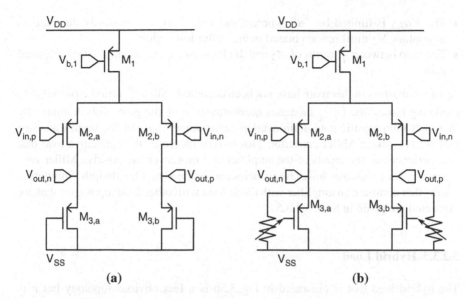

Fig. 3.5 a A unipolar differential amplifier with p-type diode-connected load transistors.
b A unipolar differential amplifier with p-type hybrid load transistors

3.2.3.2 Diode-Connected Load

The second obvious connection scheme for a p-type load transistor is the diode-connected load that is employed in the differential amplifier in Fig. 3.5a. The major advantage of the diode load is the high gate overdrive in the transistor which reduces the sensitivity of the transistor current to a ΔV_T change compared to the zero-V_{GS} load as can be seen in Fig. 3.4. Although it is clearly visible in the linear graph that the absolute current error is larger than in the zero-V_{GS} load, the logarithmic graph demonstrates that the relative error is reduced from a factor 2 in the zero-V_{GS} load to a factor 1.3 in the diode load which is more reasonable. The low ΔV_T sensitivity is also beneficial for mismatch so the offset is lower than in an amplifier with zero-V_{GS} load. This type of load is characterized by a small-signal impedance of $1/g_{m,3}$ and a high gate overdrive. The gain A_{dcl} of this amplifier with diode-connected load is given in Eq. (3.12):

$$A_{dcl} = \frac{g_{m,2}}{g_{m,3}} = \frac{V_{SG,3} - V_T}{V_{SG,2} - V_T} = \sqrt{\frac{\frac{W_2}{L_2}}{\frac{W_3}{L_3}}} \qquad (3.12)$$

The gain is determined by the ratio of the gate overdrive voltages of the transistors M_3 and M_2. The gain of an amplifier with diode load is for two reasons intrinsically smaller than when a zero-V_{GS} load is applied.

- The $V_{SG,3}$ is limited by $\frac{V_{DD}}{2}$ whereas the $V_{SG,2}$ can not endlessly decrease as transistors M_2 must remain biased in the saturation region.
- The ratio between $\frac{W}{L}$ ratios of M_2 and M_3 increases quadratically with the desired gain.

As the transistors in this work have not been designed with a $\frac{W}{L}$ ratio below 5 due to modeling issues, the $\left(\frac{W}{L}\right)_2$ increases quadratically with the gain. This dramatically increases the parasitic gate-drain overlap capacitance $C_{gd,2}$ of the input transistor that behaves like a Miller capacitor. That in turn increases the capacitive load that is experienced at the inputs of the amplifier and moreover the positive Miller zero frequency shifts towards lower frequencies and in the signal bandwidth. Simulation results that compare an amplifier with diode load with other load topologies that are presented are given in Sect. 3.2.3.5.

3.2.3.3 Hybrid Load

The hybrid load that is presented in Fig. 3.5b is a less obvious topology but it is interesting from a theoretical point of view. This topology is a trade-off between the zero-V_{GS} load and the diode load. The bias voltage at the gate is now an intermediate voltage determined by the resistive divider and situated between the source voltage and the drain voltage. The properties of the hybrid load are all in a direct trade-off between the diode load with a better reliability and the zero-V_{GS} load with a higher gain. Depending on the application with this topology either a defensive or an aggressive balance can be chosen. The drawback of this topology is that it is built using resistors that are not available in the applied technologies. Moreover this topology does not meet the specification of high gain combined with high reliability. Simulation results comparing an amplifier with hybrid load with other load topologies are given in Sect. 3.2.3.5.

3.2.3.4 Bootstrapped Gain-Enhancement

The zero-V_{GS} load and the diode load each have an important advantage but they fall short in providing a load topology that scores well on high reliability and high gain at the same time. The hybrid load enables the trade-off between both advantages but also fails to combine them. A solution that effectively combines both advantages is shown in Fig. 3.6a. The bootstrapped gain-enhancement (BGE) technique employs a high-pass RC filter, implemented as a capacitor C and a switch transistor M_{bge}, together with the p-type load transistor M_3. The capacitor C is connected to the source and the gate of M_3 while the switch connects the gate to V_{SS}. The behavior of this load is explained by a precharge and a hold phase. When V_{bge} is low in the precharge phase C is charged with V_{out} and the load acts like a diode load. Next in the hold phase the V_{bge} is high and C remains charged. As a result transistor M_3 is biased with a high gate overdrive in DC, just like the diode load, hence this load scores well

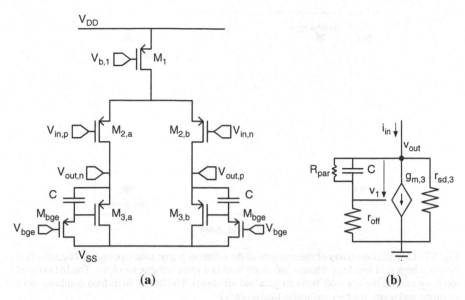

Fig. 3.6 **a** A unipolar differential amplifier with p-type load transistors and bootstrapped gain-enhancement. **b** The small-signal equivalent scheme of the BGE load

for reliability. Moreover as the $V_{SG,3}$ is fixed on the capacitor, the transconductance $g_{m,3}$ is bootstrapped out and only $r_{0,3}$ is experienced in the load transistor. Besides the good reliability of this load it also reaches a high load impedance. Both advantages are now fully combined in one p-type load topology without being in a trade-off. In Fig. 3.7 the relative strengths and weaknesses of the different p-type load topologies are summarized. The BGE load combines both strengths, i.e. high reliability and high gain, whereas the hybrid load is only a trade-off between the diode load and the zero-V_{GS} load.

The BGE load is not a trade-off between gain and reliability but a combination of both. Unfortunately, this type of load has another drawback that is caused by the high-pass filter. In the hold phase the switch M_{bge} has a high yet finite impedance r_{off}. The equivalent scheme of the load is presented in Fig. 3.6b. The impedance r_{off} together with the capacitor C and the equivalent circuit of M_3 defines the frequency dependent impedance, given in Eq. (3.13):

$$r_L = (r_{sd,3} \parallel \frac{1}{g_{m,3}}) \cdot \frac{1 + j2\pi f \cdot r_{off}C}{1 + j2\pi f \cdot \frac{(r_{off}+r_{sd,3})C}{1+g_{m,3}\cdot r_{sd,3}}}$$

$$\approx \frac{1}{g_{m,3}} \cdot \frac{1 + j2\pi f \cdot r_{off}C}{1 + j2\pi f \cdot \frac{r_{off}C}{1+g_{m,3}\cdot r_{sd,3}}} \tag{3.13}$$

At very low frequencies the impedance of the BGE load is $1/g_{m,3}$, which is just like the impedance of the diode load. Then the impedance value experiences respectively

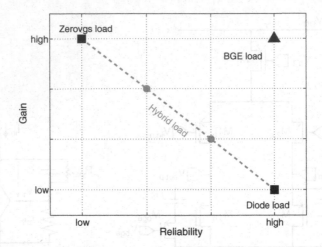

Fig. 3.7 Graphical summary of the strengths of the different p-type load topologies. The zero-V_{GS} load is a high gain topology whereas the diode load is a more reliable topology. The hybrid load topology enables the trade-off between gain and reliability. Finally the BGE load combines both strengths and seems the best applicable load topology

a zero and a pole and finally reaches its maximal impedance $r_{sd,3}$. This zero-pole doublet is preferably situated at very low frequencies, in theory at 0 Hz. In practice, however, the parasitic parallel resistance R_{par} in the capacitor determines a lower limit for this doublet. The capacitor slowly discharges through this resistor in the hold phase. The condition for keeping the gate of M_3 biased at the ground level in DC, is that $r_{off} \ll R_{par}$.

3.2.3.5 Simulation

Simulations have been carried out to compare the four presented p-type load topologies and their behavior in a differential amplifier under heavy conditions. The simulations have been subdivided in three specific cases, i.e. the case of a ΔV_T change for all the transistors, the case of V_T mismatch in the load transistors M_3 and the case of V_T mismatch in the input transistors M_2. All the transistors in the simulations are transistors with a backgate that is connected to the gate. As discussed in Sect. 2.4.3.2, this results in an 80 % increase of the transistor current and it is also beneficial for gain.

In the first case a ΔV_T change is present in all the transistors (M_1, M_2 and M_3) of the amplifier. This case represents the uncertainty of the exact value of V_T after processing and is the case with the highest relative importance in this section. The simulations have been performed for amplifiers that all have identical transistors M_1 and M_2, as in Fig. 3.6, but with the different load topologies. The simulation results are visualized in Fig. 3.8. The common-mode output voltage of all topologies is very

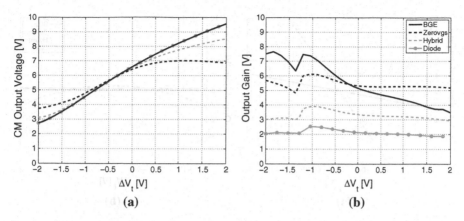

Fig. 3.8 The effect of a ΔV_T change on the common-mode output voltage and the gain of a differential amplifier for four different load topologies

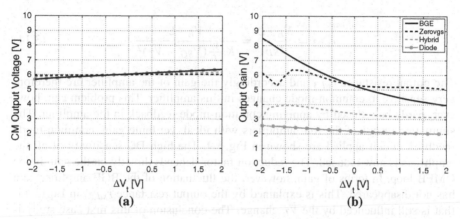

Fig. 3.9 The effect of a ΔV_T change on the common-mode output voltage and the gain of a differential amplifier with ideal, yet finite loop gain, CMFB for four different load topologies

sensitive to ΔV_T changes due to the input transistors that are biased with a small gate overdrive. This is also the reason why the zero-V_{GS} load amplifier performs better in this case where the high sensitivity of the input transistors is partly neutralized by an inverse sensitivity in the load transistors. It is clearly visible that the BGE amplifier and the diode load amplifier show an identical DC behavior as expected. As can be seen in the right panel of Fig. 3.8 the diode load, the hybrid load and the zero-V_{GS} load amplifiers perform with a stable gain that is the lowest for the diode load and the highest for the zero-V_{GS} load. Although the BGE amplifier performs with a high gain, this gain is sensitive to ΔV_T changes and drops down for high values of ΔV_T. The explanation for this remarkable observation is given by the output resistance $r_{sd,3}$ of the BGE load transistor that is inversely proportional to the squared gate overdrive, as shown in Eq. (3.14).

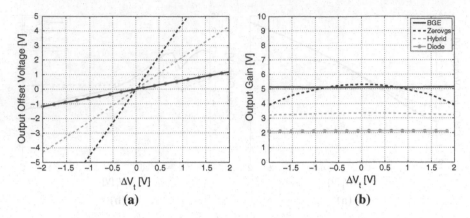

Fig. 3.10 The effect of a ΔV_T mismatch in the load transistors on the output offset voltage and the gain of a differential amplifier for four different load topologies

$$r_{sd,3} = \frac{V_E \cdot L}{I_{SD}} = \frac{V_E \cdot L}{K' \frac{W}{L}(V_{SG} - V_T)^2} \tag{3.14}$$

The output resistance $r_{sd,3}$ and accordingly the gain of the amplifier are therefore very sensitive to common-mode changes in this amplifier. This problem can, however, partly be reduced by applying common-mode feedback in the amplifier. The simulation results of the four amplifiers with ideal, yet finite gain, common-mode feedback circuit applied are shown in Fig. 3.9. The high DC sensitivity of all the amplifiers is now reduced. The reduction factor depends directly on the gain in the CMFB loop. In terms of gain, however, the fluctuation of the BGE amplifier gain has not disappeared. This is explained by the output resistance $r_{sd,3}$, in Eq. (3.14) that is still influenced by the V_T change. The conclusion of this first case study is that the BGE amplifier and the zero-V_{GS} load amplifier are the solutions that produce the highest gain and that the gain of the zero-V_{GS} load is more stable than the gain of the BGE load. The implementation of CMFB in the amplifier is further discussed in Sect. 3.3.1.1. The abnormalities present on the lines in the gain plots around $\Delta V_T = -1.5 V$ are explained by a rough transition in the transistor model between the saturation and the subthreshold region. They can be further neglected in this discussion.

The second case that is simulated is the case of a V_T mismatch present in the load transistors. With this case the sensitivity of the load topologies comes clearly to the surface and the elimination is further executed. The simulation results are shown in Fig. 3.10. In the right panel one can see that the diode-load amplifier and the BGE amplifier perform well and produce a low output offset. The zero-V_{GS} load performs poorly since the load transistors are biased with a gate overdrive close to zero. The ratio between the sensitivities of the BGE amplifier and the zero-V_{GS} amplifier is 1 : 8. The hybrid load again shows a behavior that is a clear trade-off between the diode load and the zero-V_{GS} load. The gain of the BGE amplifier is very stable and

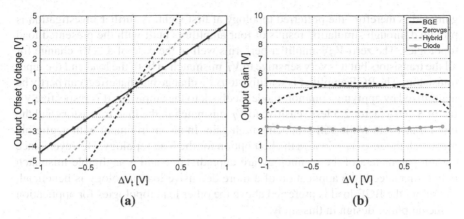

Fig. 3.11 The effect of a ΔV_T mismatch in the input transistor on the output offset voltage and the gain of a differential amplifier for four different load topologies

hardly varies whereas the zero-V_{GS} amplifier has a gain that is more sensitive to V_T mismatch in the load. The latter is of course caused by the DC offset that is generated in the zero-V_{GS} load due to this mismatch. Again the hybrid load has an intermediate performance between diode load and zero-V_{GS} load.

The third and last case that is simulated is V_T mismatch in the input transistors M_2. This case is important because the input transistors in all four topologies have a small gate overdrive and are therefore a priori very sensitive to V_T mismatch in the transistors M_2. Therefore it is the load behavior that makes or breaks each amplifier. The simulation results are shown in Fig. 3.11. The diode load and the BGE load again produce a lower offset than the hybrid load and the zero-V_{GS} load. This time the ratio between the offset of the BGE load and the zero-V_{GS} load is 1 : 2.5. The DC offset in turn influence the gain of the amplifiers. Compared to the previous case, about V_T mismatch in the load transistors, all topologies are equally or more affected by the mismatch in the input transistors yet the BGE load still performs better than the zero-V_{GS} load.

3.2.3.6 Conclusion

The demand for a p-type load topology that replaces a typical n-type load transistor forces the designer to explore alternative solutions since the obvious solutions display shortcomings. Four load topologies, i.e. the diode load, the zero-V_{GS} load, the hybrid load and the bootstrapped gain-enhancement load, have been presented and their behavior in organic electronics technology is compared with respect to the known technology-related challenges such as the V_T issues. The zero-V_{GS} load produces high gain but it is very sensitive to V_T change and mismatch whereas the diode load is a more reliable topology that produces a low gain. The hybrid load enables a trade-off between these extremes. The BGE load combines the strengths of reliability and gain and is therefore the preferred topology at first sight. A further investigation is provided through simulation results of four amplifiers built with the presented load topologies. The zero-V_{GS} amplifier performs well in the case of a ΔV_T change for all the transistors but it is very sensitive to V_T mismatch in both the load and the input transistors. The BGE amplifier is sensitive to ΔV_T change but that effect can partly be neutralized by applying common-mode feedback to the amplifier. It is respectively a factor 8 and a factor 2.5 less sensitive to V_T mismatch in the load transistors and the input transistors while still producing high gain. In an environment with negligible mismatch the zero-V_{GS} load is the best option for the load topology. However, since certain parameters of the technology are unpredictable and a negligible mismatch is not guaranteed, the application of a more defensive load topology is beneficial. Therefore the BGE load is preferred above the other load topologies for application in the amplifier design in this study.

3.2.4 Circuit Techniques

Very often applications demand an amplifier gain that is orders of magnitude higher than the gain of a single-stage differential amplifier. This high performance is reached by applying additional circuit techniques, such as cascode transistors and gain boosting. The application of these techniques in organic electronics technology in a reliable way is often a challenge and calls for caution. In this section these techniques are discussed and their applicability is investigated.

3.2.4.1 Cascoded Amplifier

The schematic view of a differential amplifier with cascode transistors is presented in Fig. 3.12a. The gates of the cascode transistors M_4 are biased with the external voltage V_{casc}. The gain of this amplifier is calculated in Eq. (3.15).

$$A_{casc} = -g_{m,2} \cdot \left(r_{sd,3} \parallel (1 + g_{m,4} \cdot r_{sd,4}) \cdot r_{sd,2} \right) \tag{3.15}$$

This equation proves that only the impedance experienced at the output node towards the cascode transistor is increased whereas the impedance of the load transistor remains identical. The effect of the cascode transistor on the gain of the amplifier is therefore small. Moreover the cascode transistor is difficult to bias and increases the V_T sensitivity of the amplifier, both in terms of V_T change and V_T mismatch, since it is biased with a small gate overdrive for high gain. Yet the cascode transistor is applicable for increasing the impedance of transistor M_1. This impedance must be very high compared to the $\frac{1}{g_{m,2}}$ impedance experienced towards the input transistors M_2 in order to increase the CMRR. The cascoded current transistor is further discussed in Sect. 3.3.1.2.

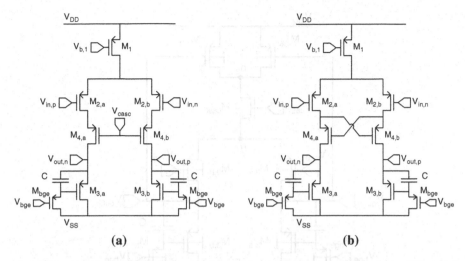

Fig. 3.12 a A unipolar differential amplifier with cascode transistors. b A unipolar differential amplifier with gain boosting

3.2.4.2 Gain Boosting

A technique that increases the effect of a cascode transistor is gain boosting: instead of biasing the gate of the cascode transistor $M_{4,a}$ and $M_{4,b}$ with a DC voltage only, a signal opposite to the signal at the source of $M_{4,a}$ is applied. This can be done with the signal at the source of transistor $M_{4,b}$, as shown in Fig. 3.12b. This circuit wrongfully appears to apply positive feedback, yet the technique only applies feed forward that doubles the small-signal $v_{sg,4}$ which is experienced. This results in the gain A_{gb} given by Eq. (3.16):

$$A_{gb} = -g_{m,2} \cdot \left(r_{sd,3} \parallel (1 + 2 \cdot g_{m,4} \cdot r_{sd,4}) \cdot r_{sd,2} \right) \tag{3.16}$$

The considerations about this topology are almost identical to those of the cascoded amplifier. Although this topology is self-biased, it is very sensitive to V_T change and V_T mismatch. Moreover the effect on the gain of the amplifier is limited by the impedance of the load.

Both the cascoded amplifier and the amplifier with gain boosting are amplifiers, which are very sensitive to V_T change and mismatch and which can only partly increase the gain of the amplifiers. The best way to increase the gain of an amplifier is to produce a cascaded n-stage amplifier. This technique is applied and further discussed in Sect. 3.3.2.

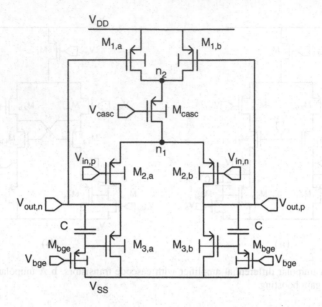

Fig. 3.13 Schematic view of the single-stage amplifier with linear CMFB, cascoded current source, GBG steering and BGE load

3.3 Designs

After a profound discussion in Sect. 3.2 a differential amplifier with BGE load has been selected as the best starting point in amplifier design. In this section the design and the measurement results of a single-stage amplifier are discussed. Consequently a three-stage amplifier is presented that employs high-pass filters to connect the consecutive stages. Then an improved amplifier design is presented that enables the DC connection between consecutive stages. Finally a comparator, based on the first single-stage amplifier design is proposed.

3.3.1 Single-Stage Amplifier

The implementation of the single-stage amplifier is presented in Fig. 3.13. This amplifier is built with 9 transistors and applies linear common-mode feedback, a cascoded current source, gate-backgate steering in all transistors and bootstrapped gain-enhancement (BGE) to improve the amplifier specifications and to overcome processing and behavioral variations in the transistors. Every of these techniques is discussed in Sects. 3.2.3.4–3.3.1.1. Consequently simulation results are shown in Sect. 3.3.1.5 that investigate the behavior of the circuit under heavy conditions.

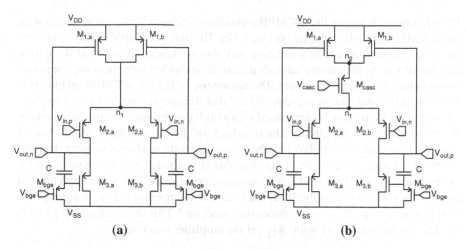

Fig. 3.14 Part by part build-up of the single-stage differential amplifier. **a** The amplifier with BGE load and CMFB. **b** The amplifier with BGE load, CMFB and a cascoded current source

Finally the measurement results of the single-stage amplifier are presented and discussed in Sect. 3.3.1.6.

3.3.1.1 Linear Common-Mode Feedback

At the moment, after a thorough discussion, the p-type load transistors M_3 are implemented as BGE loads and have become very insensitive to V_T changes in the load transistors. Conversely, the input transistors M_2 are biased with a low gate overdrive close to subthreshold for high gain. This low gate overdrive is actually forced by the definition of the word 'amplifier', i.e. that an amplifier amplifies signals with a gain larger than 1 at least, and is therefore not contested. As the intrinsic gain of a transistor is limited there is no abundance of gain available to exchange for reliability. Unfortunately the low gate overdrive increases the sensitivity of the amplifier to V_T change and V_T mismatch in transistors M_2. Simulation results that confirm this sensitivity are given in Sect. 3.2.3.5. In the situation of an identical ΔV_T change in both transistors M_2 this has an effect that is inverse but identical to a common-mode voltage at the input ΔV_{in}. It is therefore logical that a CMFB loop arms the amplifier against such an effect. Against the other effect, V_T mismatch in M_2, on the other hand, no techniques can be applied except for offset compensation. Offset compensation is added to this topology in the three-stage op amp in Sect. 3.3.2. The subsequent single-stage amplifiers are connected through high-pass filters that compensate for the offset of each state.

The implementation of common-mode feedback is hampered by the unipolar character of the technology. Therefore several CMFB architectures are a priori discarded as they require n-type as well as p-type transistors. The easiest way to implement

CMFB is the well-known linear CMFB architecture (Sansen 2008). An amplifier with linear CMFB applied is presented in Fig. 3.14a. This technique applies two transistors M_1 biased in the linear region that convert common-mode deviations at the output in a linear way to the current through the amplifier which influences the common-mode output deviations negatively. The advantage of the linear CMFB architecture is that no additional circuits are required and that the circuit complexity remains low. Moreover the transistors M_1 are biased with a high gate overdrive which is beneficial for the reliability of the circuit. The drawback of the linear CMFB architecture is that the gain-bandwidth (GBW) of the common mode is always smaller than the GBW of the differential mode. Since the CMFB is applied here to overcome ΔV_T change which is a DC effect, this does not interfere with the aim of the implemented circuit. On the other hand the BGE load has a low impedance $1/g_{m,3}$ at very low frequencies, i.e. in DC, which reduces the effect on VTSR drastically. In Eq. (3.17) the DC common-mode V_T gain, A_{V_T}, of the amplifier is derived:

$$A_{V_T} = -\frac{1}{g_{m,3} \cdot r_{sd,1} \cdot (1 + LG)} = -\frac{1}{g_{m,3} \cdot r_{sd,1} \cdot \left(1 + \frac{g_{m,1}}{g_{m,3}}\right)} \tag{3.17}$$

In this formula $g_{m,3}$ represents the impedance at frequency 0 Hz of the BGE load. Consequently, and since M_1 is in the linear region, the loop gain (LG) is smaller than 1. Therefore the effect of the CMFB loop on the VTSR of the amplifier is a factor between 1 and 2. This is a modest but positive result.

3.3.1.2 Cascoded Current Source

In the amplifier in Fig. 3.14b a cascode transistor M_{casc} is added. This transistor is placed with the purpose to increase the impedance of the current source at the virtual ground node n_1 and accordingly to improve the common-mode behavior of the amplifier. As can be seen in Eq. (3.17), the impedance of the current source has a direct impact on the VTSR. Equation (3.18) shows the A_{V_T} of the amplifier when a cascoded current source is added.

$$A_{V_T} = -\frac{1}{g_{m,3} \cdot r_{sd,1} \cdot (1 + g_{m,c} \cdot r_{sd,c}) \cdot \left(1 + \frac{g_{m,1}}{g_{m,3}}\right)} \tag{3.18}$$

where $g_{m,c}$ and $r_{sd,c}$ are the transconductance and the output resistance of the cascode transistor M_{casc}. According to the definition of VTSR in Eq. (3.5) the cascode amplifier increases the VTSR of the amplifier.

3.3.1.3 Gate-Backgate Steering

In Sect. 2.4.3.2 the technique of backgate steering is introduced and several implementation methods are discussed. In one of those, gate-backgate steering (GBG steering), the gate and the backgate of a transistor are connected together which increases the current characteristics of that transistor with $\sim$80 %. Accordingly the transconductance g_m of the transistor increases with the same amount whereas the parasitic capacitance only increases with $\sim$25 %. Therefore the GBW is increased with $\sim$44 %. All the transistors in the amplifier are implemented with the gate-backgate steering technique as can be seen in Fig. 3.13.

3.3.1.4 Bootstrapped Gain Enhancement

The BGE load has been extensively discussed in Sect. 3.2. It is implemented with a metal-metal capacitor C_{BGE} and a transistor M_{BGE}. It creates a pole-zero doublet at a low frequency. The lower limit for this doublet is determined by the leakage in C_{BGE} through the parasitic resistance present in C_{BGE}. Measurements have revealed that the leakage of this capacitor has an RC discharging constant of 15 s, so the technology related pole of this capacitor is located at $\sim$0.01 Hz. The bump in the phase domain that is caused by the zero-pole doublet is determined by the distance between the zero and the pole. According to Eq. (3.13) this distance is determined by $g_{m,3} \cdot r_{sd,3}$ of the load transistor and difficult to predict.

3.3.1.5 Simulation

Simulations are performed to estimate the behavior of the amplifier. The transistor sizing of the amplifier is presented in Fig. 3.15 and the simulated DC operation points are included. The simulated gain of this amplifier is 4.4 or 13 dB. The simulation results have to be interpreted carefully since the simulations are performed with preceding measurement results of transistors on previous runs. Nevertheless they give an idea of what can be expected from this amplifier. In Fig. 3.16 the simulated amplifier behavior is presented when a ΔV_T change is present. These results reveal a behavior which is worse than the simulation results in Fig. 3.9. The explanation for this is that the common-mode feedback is not ideal because the loop gain is low.

3.3.1.6 Measurement Results

The single-stage amplifier has been implemented on top of a plastic foil and is measured in a nitrogen environment at room temperature to prevent the occurrence of degradation effects. The Bode plot of the amplifier is presented in Fig. 3.17a. It has a gain of 15 dB and a GBW of 10 kHz. The other curve on the graph corresponds to the same amplifier in the identical measurement setup, yet without the backgates on

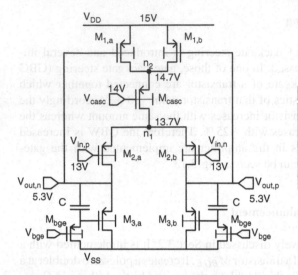

Component	Size
M_1	2.1 mm/5 µm
M_2	3.7 mm/5 µm
M_3	300 µm/5 µm
M_{casc}	11.7 mm/5 µm
M_{bge}	140 µm/130 µm
C	34 pF

Fig. 3.15 Simulated operating points of the differential amplifier and a list of the transistor sizes

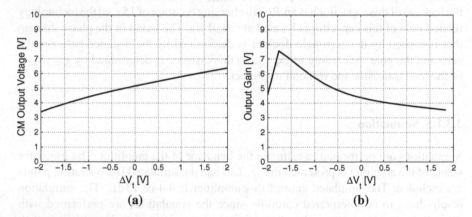

(a)

(b)

Fig. 3.16 The simulated effect of a ΔV_T change on the common-mode output voltage and the gain of the presented differential amplifier

top of the transistors. This amplifier has a gain of only 8 dB and a GBW of 3.3 kHz. The measured factor 3 between the GBW with and without backgates is higher than the expected factor 1.44 but the marked 7 dB difference of the gain is pretty well explained by the factor 2.4 (7.6 dB) predicted in Sect. 2.4.3.2. The difference of the GBW is likely explained by a shift in the DC operating points of the amplifier caused by the virtual V_T change in the transistors caused by the gate-backgate steering technique. The measured CMRR of the amplifier is 12 dB. The VTSR is the ratio between the gain of the amplifier in the signal bandwidth and the common-mode gain at 0 Hz and is therefore larger than the CMRR. The amplifier consumes 1.5 µA

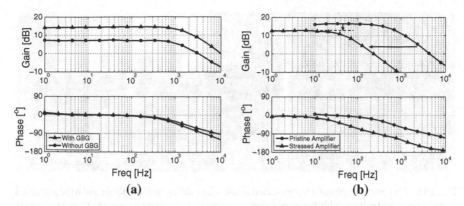

Fig. 3.17 **a** The measured Bode plot of the single-stage differential amplifier. **b** The measured Bode plot of pristine amplifier and the measured degradation of the amplifier after four months shelving followed by 10 days biasing at 15 V

from a V_{DD} of 15 V. The area of this amplifier measures $2 \times 2 \text{ mm}^2$. As discussed in Sect. 3.3.1.4 the BGE load introduces a zero-pole doublet at low frequencies. The measurements have shown that the lower boundary for this doublet is situated at 0.1 Hz. This limit is originated from the parasitic parallel resistance in the capacitor.

The long-term behavior of this amplifier in ambient environment at room temperature is measured. The results are shown in Fig. 3.17b, where the measured Bode plot of the pristine amplifier and the Bode plot of the amplifier after four months shelving followed by 10 days biasing at 15 V are visualized. Two distinct trends show up.

- There is a decrease of the gain, which amounts to 4 dB,
- The GBW decreases from 4 kHz to 300 Hz.

Both effects are caused by a shift of the V_T and the mobility μ in the transistors of the amplifier, caused by bias stress and O_2 and humidity in the air.

This amplifier is carefully designed for reliable DC behavior and high gain. Several techniques have been applied that improve both the reliability and the gain. Table 3.1 summarizes the measurement results of this amplifier. The chip photograph is presented in Fig. 3.18a.

3.3.2 3-Stage Operational Amplifier

An ideal op amp has an infinite gain and an infinite input resistance. It is a building block that is often used in feedback where its gain determines the precision of the circuit: the higher the gain is, the better the circuit performance. The gain of a single-stage amplifier is often not high enough to meet the specifications given by an application. In that case a cascaded amplifier with more stages is required. In this section a three-stage operational amplifier is presented that is built up with three

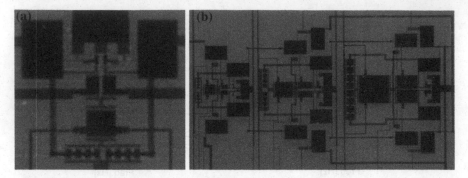

Fig. 3.18 Chip photographs of **a** the presented single-stage differential amplifier and **b** the presented 3-stage op amp built up with the single-stage amplifier. These circuits are made in technology provided by Polymer Vision

Table 3.1 Summary of the measurement results for the single-stage differential amplifier integrated on plastic foil (Marien et al. 2010, 2011).

Specification	Value
Power supply	15 V
Gain	15 dB
Band width	0.6 kHz
GBW	10 kHz
CMRR	12 dB
VTSR	>12 dB
Current consumption	1.5 μA
Power	15 μA
Chip area	$2 \times 2\,\text{mm}^2$

single-stage amplifiers presented in Sect. 3.3.1. In Sect. 3.3.2.1 the implementation of the op amp is presented. Finally the measurement results of the op amp are presented in Sect. 3.3.2.2.

3.3.2.1 Implementation

The implementation of the 3-stage op amp is presented in Fig. 3.19. It is built up with three single-stage amplifiers. The first one is identical to the amplifier discussed in Sect. 3.3.1, whereas the second and the third amplifier are enlarged with a factor 3 and 9 respectively. As the operating points of the input and the output nodes in the single-stage amplifier are different, a level shifter is required in between each two consecutive amplifiers. In this design a passive implementation for this level shifter, a high-pass filter, is applied. The passive implementation is preferred over a level shifter based on a source follower, since the maximal simulated AC throughput in a level shifter built with p-type organic transistors amounts to ∼60 % whereas the losses in the passive implementation are negligible and the optimal ∼100 % throughput is reached. The passive implementation is identical to the high-pass filter

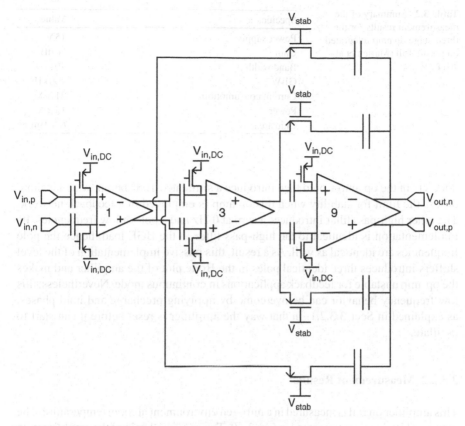

Fig. 3.19 Implementation of the presented 3-stage op amp

applied in the BGE load and employs a 34 pF capacitor and a 25 μm/5 μm switch that determines if the op amp is in the precharge phase or in the hold phase.

In order to stabilize the op amp, pole splitting is applied through Miller capacitors which have a 34 pF capacitance. Additionally, feedforward blocking resistors, implemented with 1.5 mm/15 μm transistors biased in the linear region, are included. Contrarily to the high-frequency behavior which is stabilized, the low-frequency behavior is not stable in the continuous mode. It requires a precharge phase that resets the starting conditions for the amplifier.

In the design of this amplifier there are two types of high-pass filters implemented.

- The high-pass filter that is present in the BGE load,
- The high-pass filter that is used as a level shifter in between the consecutive stages.

Their different function results in a different effect on the low-frequency behavior of the amplifier. The high-pass filter in the BGE load introduces a zero-pole doublet in the amplifier behavior as mentioned in Sect. 3.3.1.4, hence a simulated 36° bump is present in the phase characteristics of the amplifier. The placement of three identical

Table 3.2 Summary of the measurement results for the three-stage op amp integrated on plastic foil (Marien et al. 2011)

Specification	Value
Power supply	15 V
Gain	26 dB
Band width	40 kHz
GBW	700 kHz
Current consumption	21 μA
Power	15 μA
Chip area	$6 \times 9 \, mm^2$

doublets in the op amp would then introduce a simulated 108° bump, which is more or less the limit for stability when this op amp is employed in a feedback network. The other high-pass filter introduces a zero at 0 Hz and a pole at low frequencies. Its implementation is identical to the high-pass filter in the BGE load, hence the pole frequencies are identical as well. As a result, this passive implementation of the level shifters introduces three identical poles in the Bode plot of the amplifier and makes the op amp unstable for feedback applications in continuous mode. Nevertheless, this low-frequency behavior can be overcome by applying precharge and hold phases, as explained in Sect. 3.3.2.1. In that way the amplifier is reset before it can start to oscillate.

3.3.2.2 Measurement Results

This amplifier on foil is measured in a nitrogen environment at room temperature. The measured Bode plot is presented in Fig. 3.20. The measured gain of the amplifier is 26 dB and the GBW amounts to 700 Hz. The gain is 15 dB lower than what is expected from the behavior of the single-stage amplifier. This is explained by the high-pass filters that have not been scaled up together with the single stages. The requirement for the high-pass filter throughputs between the stages is that the input capacitance C_{in} needs to be much smaller than the capacitance of the filter, C_f. If not, a capacitive division reduces the throughput of the filter. This is no more the case for the input capacitance of the second and especially the third stage and in this way the gain of the op amp is unfortunately decreased. This problem can easily be fixed by sizing the high-pass filters together with the single-stage amplifier that it is meant to drive. The phase margin (PM) of this amplifier for high frequencies is 70°. The op amp consumes 21 μA from a 15 V power supply voltage.

This op amp has been designed for high gain and for use in other analog circuits. The implementation on foil measures $6 \times 9 \, mm^2$. The measurement results are summarized in Table 3.2. The chip photograph is shown in Fig. 3.18b.

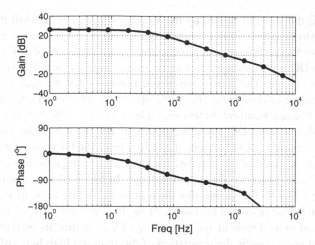

Fig. 3.20 The measured Bode plot of the presented 3-stage op amp

3.3.3 Improved Amplifier

The single-stage amplifier that is presented in Sect. 3.3.1 is optimized for high gain in a single stage. All techniques were applied so as to maximize the gain. The usage of this single-stage amplifier in the 3-stage op amp, discussed in Sect. 3.3.2 requires level shifters that overcome the DC voltage gap between the output and the input operating points. The level shifters, implemented with passive high-pass filters, cause stability problems at low frequencies when the op amp is employed in a feedback network. Therefore a single-stage amplifier with identical input and output operating points, that no longer needs a level shifter, is of interest and justifies a certain sacrifice in terms of gain. In this section an improved DC-coupled amplifier is presented and the techniques that are required are discussed.

3.3.3.1 Threshold Voltage Tuning

The influence of the backgate voltage of a transistor is a linear V_T shift of that transistor, as explained extensively in Sect. 2.4.3.1. This technique applied in a transistor enables to shift the gate voltage together with V_T, while the transistor current remains unchanged, as explained in Eq. (3.19):

$$I_{SD} = K'_p \cdot \frac{W}{L}(V_{SG} - V_T)^2$$
$$= K'_p \cdot \frac{W}{L}((V_{SG} + \Delta V_{SG}) - (V_T + \Delta V_T))^2 \qquad (3.19)$$

where ΔV_T is the V_T shift caused by the V_{SBG} and ΔV_{SG} is the shift that the V_{SG} must undergo to keep the transistor current I_{SD} identical. ΔV_{SG} is of course identical to ΔV_T. When this technique is applied to the input transistors M_2 of a differential amplifier the DC input voltage V_G can be shifted up or down, depending on the backgate bias voltage. The conclusion of this theory is that the backgate enables a degree of freedom to change the input DC operating point of an amplifier without changing the transistor current. In this way it becomes possible to bias a differential amplifier with the same DC operating points at the input and the output nodes. Knowing the ratio of ~ 0.33 between the influence of the backgate and the gate, an estimation can be made of the required backgate bias voltage to bias the input DC operating points identical to the output voltage. In Fig. 3.13 the required V_{SBG} is $-3.\Delta V_{SG}$ or -23.1 V. This means that the required absolute value of V_{tvt} is around 38 V. The exact value of course also depends on the transistor sizing. This technique will be referred to as threshold voltage tuning (TVT) in this dissertation. The new challenge that arises now is the generation of the required high bias voltage of 30–40 V. This challenge is discussed and solutions for this challenge are presented in Chap. 6.

3.3.3.2 Implementation

The implementation of a 2-stage DC-coupled operational amplifier is presented in Fig. 3.21. Each single-stage amplifier is built with an input transistor pair M_2 that is biased through the V_{tvt} pin. Furthermore the amplifier applies BGE load and CMFB, just like in the previous designs. The DC operating points of the amplifier are included in the figure. As can be seen, the input and the output operating points are identical. This is obtained at the expense of a high bias voltage V_{tvt} that amounts to 35 V. The backgates of the current transistors M_1 are also biased with the same bias voltage V_{tvt}. Therefore M_1 are no more biased in the linear region but in saturation region. This increases the gain in the CMFB loop as well as the output impedance of the current source, which makes the previously used cascode transistor obsolete. The drawback, however, of the current sources biased in saturation, is that the CMFB is no longer linear and that the differential signals are no longer linearly cancelled. This can result in a reduction of the differential gain.

3.3.3.3 Measurement Results

The measured Bode plots of a 1-, a 2- and a 3-stage DC connected amplifier are presented in Fig. 3.22a. The gain of the amplifiers is proportional to the number of stages as expected. The gain of a single-stage amplifier, which is around 10 dB, is lower than the gain of the amplifier presented in 3.17. This is of course explained by the different technique that is applied to employ the backgate. In that amplifier discussed in Sect. 3.3.1 the gate and the backgate of the input transistor are connected, which improves $g_{m,2}$ and accordingly the gain, whereas in the amplifier presented in

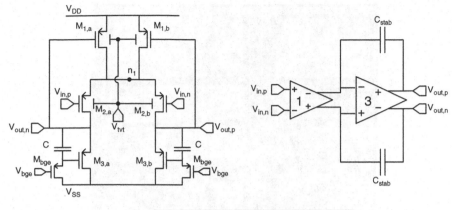

Component	Size
M_1	20 mm/5 µm
M_2	20 mm/5 µm
M_3	100 mm/5 µm
M_{bge}	140 mm/130 µm
C	34 pF
C_{stab}	8 pF

Fig. 3.21 Implementation of the presented improved single-stage amplifier and the DC-coupled 2-stage op amp and a list of the component sizes

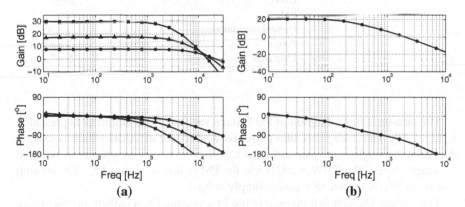

Fig. 3.22 The measured Bode plots of **a** a 1-, 2- and 3-stage DC connected amplifier and **b** of the DC-coupled 2-stage op amp

this section the TVT technique is applied to equalize the input and output operating points, which of course does not increase $g_{m,2}$ and the gain. The 5 dB difference between the gain of the two amplifiers is explained by the 80 % of the $g_{m,2}$ when GBG steering is applied, as in the other design.

Fig. 3.23 Chip photograph of the presented DC-coupled two-stage op amp on foil. This circuit is made in technology provided by Polymer Vision

Table 3.3 Summary of the measurement results for the two-stage op amp integrated on plastic foil (Marien et al. 2011)

Specification	Value
Power supply	15 V
Gain	20 dB
Band width	200 kHz
GBW	2 kHz
PM	65°
Current consumption	15 μA
Power	225 μW
Chip area	$2 \times 2.4 \, \text{mm}^2$

The measured Bode diagram of the 2-stage version of this amplifier is presented in Fig. 3.22b. The 2-stage op amp performs with a gain of 20 dB which is exactly double of the single-stage gain as in this topology no losses are present in between the single stages. The GBW is 2 kHz and the PM of this op amp is 65°. The op amp consumes 51 μA from a 15 V power supply voltage.

The 2-stage DC-coupled op amp is the first organic DC-coupled analog circuit on foil that has been measured and reported in literature. This technique enables the generation of high gain in cascaded amplifiers and the design of larger analog circuits in a reliable and performant way. The lay-out of the 2-stage op amp measures $2 \times 2.4 \, \text{mm}^2$. The measurement results of both the single-stage and the 2-stage amplifiers are summarized in Table 3.3. The chip photograph is shown in Fig. 3.23.

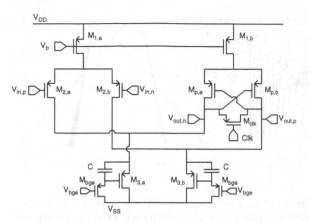

Fig. 3.24 Implementation of the presented comparator on foil and list of the component sizes

Component	Size
$M_{1,a}$	6 mm/5 μm
$M_{1,b}$	12 mm/5 μm
M_2	1.8 mm/5 μm
M_3	300 μm/5 μm
M_p	360 μm/5 μm
M_{clk}	300 μm/5 μm
M_{bge}	140 μm/5 μm
C	34 pF

3.3.4 Comparator

3.3.4.1 Implementation

The implementation of the comparator is presented in Fig. 3.24. It is built with a differential input pair M_2 and with a cross-coupled transistor pair M_p. Both transistors M_2 and M_p share the same BGE load transistor pair M_3. A clocked transistor neutralizes the cross-coupled transistors and introduces a track phase, when the clock signal is low, and a hold phase, when the clock signal is high.

3.3.4.2 Measurement Results

The comparator circuit consumes 9 μA from a supply voltage of 20 V. In the track phase both outputs are brought towards each other. The ΔV_{in} of the input pair determines which output node will be slightly higher. When the clock turns high, the cross-coupled pair separates both outputs. Figure 3.25a shows the output signals of the comparator for a differential triangular 2 V_{ptp} input. The gain reaches 12 dB at the trip point. This is when the BGE is activated. In the case where BGE is switched to non-active we only measure a gain of −3 dB. The upper limit for correct comparator behavior of the clock frequency is limited to 2 kHz. The lower limit is determined by the AC-coupling and below 0.01 Hz. Measurements have been done in ambient environment and at room temperature. Since the circuit is designed to be V_T insensitive, it can get round possible V_T shifts caused by ambient. The circuits remain operational even after a month of exposure to ambient.

Figure 3.25b shows an eye diagram of the sampled comparator output, $V_{out,p}$ − $V_{out,n}$ and the input signal, $V_{in,p} - V_{in,n}$, at clock frequency 1 kHz and input frequency 10 Hz. The output is clipped and plotted against the input waveform. From this figure

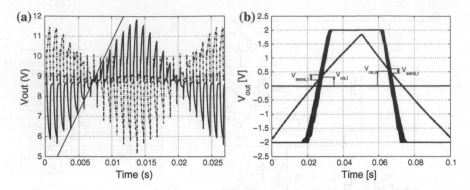

Fig. 3.25 **a** Output waveform of the comparator for a triangular 40 Hz $2V_{ptp}$ input at 1 kHz clock frequency. **b** Eye diagram of the comparator for a triangular 40 Hz input at 1 kHz clock frequency

Fig. 3.26 Photograph of a plastic foil with two comparator circuits facing eachother. This circuit is made in technology provided by Polymer Vision

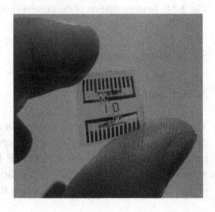

input sensitivity V_{sens}, input offset voltage V_{os} and hysteresis effects V_{hyst} can be derived, according to Eqs. (3.20), (3.21) and (3.22).

$$V_{sens} = \max(V_{sens,l}, V_{sens,r}) \tag{3.20}$$

$$V_{os} = \frac{V_{os,l} + V_{os,r}}{2} \tag{3.21}$$

$$V_{hyst} = V_{os,r} - V_{os,l} \tag{3.22}$$

Input sensitivity of this circuit results in 200 mV. Input offset is calculated as 400 mV. Finally a hysteresis of 200 mV is observed. The measurement of several samples shows that a small variation of only 2 dB on the comparator gain is present between the measured samples. These measurements demonstrate that the design strategy for low V_T sensitivity is effective. A photograph of a plastic chip with two comparators is shown in Fig. 3.26. The measurement results of the comparator are summarized in Table 3.4.

Table 3.4 Summary of the measurement results for the comparator integrated on plastic foil (Marien et al. 2009)

Specification	Value
Power supply	20 V
Gain	12 dB
Band width	1 kHz
Current consumption	9 μA
Power	180 μW
Chip area	$2 \times 3\,\text{mm}^2$

3.4 Conclusion

The first step in the exploration of analog circuits design in organic electronics was the design of an amplifier. This amplifier must be adapted to the limits and the challenges but also to the advantages and the available tricks in the technology. In the case of organic electronics technology on foil this means that only p-type transistors are available and that special care is required for variations of behavioral transistor parameters such as V_T whereas the backgate of the transistors can be advantageously employed to increase the circuit performance. From the knowledge about the technology provided in Chap. 2 a carefully considered design of an amplifier can be made.

In this chapter first the application field of organic amplifiers was highlighted. They are required for more complex analog and mixed-signal circuits such as ADCs and sensor interfaces. As an amplifier should be usable in a variety of applications no direct specifications were deducted: the gain and the bandwidth should generally be as high as possible.

Consequently the best amplifier topology was searched for. A differential amplifier was preferred over a single-ended amplifier on the basis of the threshold voltage suppression ratio. Furthermore the bootstrapped gain-enhancement load was presented as the best p-type transistor solution -better than a zero-V_{GS} load, a diode load or a hybrid load- to replace the unavailable n-type load transistors.

Then several amplifier designs were presented and their measurement results were discussed:

First, a single-stage amplifier was presented with a gain of 15 dB. This single-stage amplifier was consequently employed together with high-pass level shifters in a 3-stage op amp.

Then an improved amplifier was presented that has identical DC operating points for the input and output voltages and that can as a result easily be employed in more-stage amplifiers. This was demonstrated with a 2-stage op amp that has a gain of 20 dB. This op amp was the first DC-coupled more-stage amplifier reported in literature.

Finally a classic comparator was constructed from a basic differential amplifier. The single-stage comparator had a gain of 12 dB and performed at a clock speed of 1 kHz.

In this chapter insight was gained about reliable amplifier design and it was applied to the presented amplifiers. These were optimized and will contribute to the design of larger circuits in the following chapters.

References

Kane MG, Campi J, Hammond MS, Cuomo FP, Greening B, Sheraw CD, Nichols JA, Gundlach DJ, Huang JR, Kuo CC, Jia L, Klauk H, Jackson TN (2000) Analog and digital circuits using organic thin-film transistors on polyester substrates. IEEE Electron Device Lett 21(11):534–536

Gay N, Fischer W-J (2007) OFET-based analog circuits for microsystems and RFID-Sensor transponders. In: 6th International conference on polymers and adhesives in microelectronics and photonics polytronic 2007, pp 143–148.

Myny K, Beenhakkers MJ, van Aerle NAJM, Gelinck GH, Genoe J, Dehaene W, Heremans P (2011) Unipolar organic transistor circuits made robust by Dual-Gate technology. IEEE J Solid State Circuits 46(5):1223–1230

Sansen WMC (2008) Analog design essentials. Springer, Berlin

Marien H, Steyaert M, van Veenendaal E, Heremans P (2010) Analog techniques for reliable organic circuit design on foil applied to an 18 dB single-stage differential amplifier. Org Electron 11(8):1357–1362

Marien H, Steyaert MSJ, van Veenendaal E, Heremans P (2011) A fully integrated ΔΣ ADC in organic thin-film transistor technology on flexible plastic foil. IEEE J Solid State Circuits 46(1):276–284

Marien H, Steyaert M, van Veenendaal E, Heremans P (2011) DC-DC converter assisted two-stage amplifier in organic thin-film transistor technology on foil. Proceedings of the ESSCIRC, In, pp 411–414

Marien H, Steyaert M, van Aerle N, Heremans P (2009) A mixed-signal organic 1kHz comparator with low VT sensitivity on flexible plastic substrate. In: Proceedings of ESSCIRC (ESSCIRC'09), pp 120–123.

Chapter 4
A/D Conversion

Analog-to-digital converters (ADCs) and digital-to-analog converters (DAC) are key elements in the world of electronic circuits. They provide the interfaces between the digital processing unit and the analog world. ADCs and DACs are required in every circuit technology and exists in several architectural flavours. The interest for organic ADCs originates from the future application of organic smart sensor systems which combine sensors, actuators and a digital processing unit on a plastic foil in a biomedical, logistic or other environment. Nevertheless the implementation of ADCs in organic electronics technology is at the moment hampered by the technological challenges and requires special attention and well-considered design decisions.

The proposed architecture of an organic smart sensor system is shown in Fig. 1.5. The subject of this chapter is the design and implementation of the third building block, i.e. organic ADCs on foil. The application field is discussed in Sect. 4.1 where also the design specifications are deducted. In Sect. 4.2 a comparative study is presented which investigates the applicability of the existing ADC architectures. Section 4.3 discusses the design of two preferred architectures, a 1st and a 2nd order ΔΣ ADC. Section 4.4 brings a more profound discussion about the presented results and the applicability of the circuits. Finally the chapter is concluded in Sect. 4.5.

4.1 Application Field

One of the expected applications of organic electronics are organic smart sensor systems: A digital RFID tag on foil calculates, stores and communicates information gathered through certain sensors, e.g. temperature, chemical or pressure sensors. Typically these sensors collect analog data from the surrounding analog world and therefore also generate an analog output signal. As soon as organic smart sensor systems need to interface with organic sensors (and actuators), the demand for ADCs (and DACs) integrated in organic technology rises to the surface. These ADCs convert

H. Marien et al., *Analog Organic Electronics*, Analog Circuits and Signal Processing, DOI: 10.1007/978-1-4614-3421-4_4, © Springer Science+Business Media New York 2013

the amplified analog sensor output to a digital word that can be understood and dealt with in the digital core of the system.

Organic smart sensor systems find applications in the food industry, in warehouse logistics and in human health monitoring where the signals that are to be sensed typically fluctuate at frequencies below 10 Hz. The small bandwidth explains the good applicability of these slow systems in the organic technology on foil.

The specifications for the ADC for application in smart sensor systems are now derived from the application field. A signal bandwidth of 10 Hz in the ADC is explained to be sufficient for dealing with the slow sensed signals. This is indeed low but in the future, when the technology has improved, the speed and the bandwidth will raise automatically in later designs.

In terms of accuracy it is difficult to put forward a certain specification since this specification depends on the goal of the application. The accuracy is furthermore limited by the technological parameters, such as mismatch. Therefore in this dissertation the accuracy is just optimized. The same holds for the power consumption in the ADC yet with a lower priority. Wherever a trade-off between accuracy and power is present, more importance is assigned to accuracy.

4.2 ADC Architecture

There are tens of different ADC architectures that all differ in the way they convert the analog signals into digital signals. As a consequence they all manage the present challenges such as mismatch, V_T change and bias stress in a different manner. Some architectures depend on large banks of matched components. These suffer from the mismatch present in organic electronics technology. Other architectures are more sensitive to V_T change or bias stress. Moreover all the architectures differ among themselves in their overall performance. Some architectures perform better in terms of accuracy whereas others score better for speed. In this section the different sub-families of ADCs are presented and their qualities are discussed. Their strong points and their weaknesses are considered in order to find the optimal architecture that performs best in the given organic electronics technology (Hagen et al. 2011).

4.2.1 Flash ADC

Flash ADCs are the fastest amongst the ADCs. An n-bit flash ADC, presented in Fig. 4.1a is built up with a chain of 2^n matched resistors. They are all connected in series and behave like a voltage divider that generates $2^n - 1$ equidistant digital decision levels. The matching properties of the resistors determine the absolute and relative errors that are present in the digital levels. Furthermore $2^n - 1$ comparators are used that compare the analog input signal with each of the digital decision levels. The output of this architecture is a thermometer-code digital signal that can be digitally

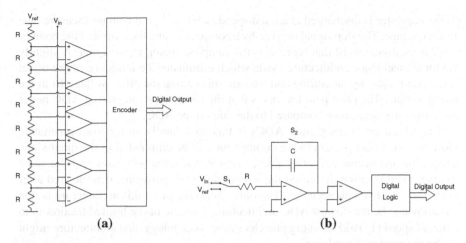

(a)   **(b)**

Fig. 4.1 Schematic system level diagrams of **a** the flash ADC architecture and **b** the integrating ADC architecture

converted to a binary word. Due to the parallellism in this architecture, flash is the fastest ADC architecture. In each clock cycle one word can be converted. On the other hand the accuracy of this type of converters is low, not only because of mismatch, but also because the area scales exponentially with the required accuracy.

The applicability of flash ADC in organic electronics technology is poor. This architecture relies on a bank of 2^n matched resistors, which are not available in the organic technology and which have to be replaced with transistors biased in the linear region. The latter are mismatch sensitive and jeopardize the accuracy of the converter. Furthermore, $2^n - 1$ comparators are required with a high gain. The high gain is difficult to reach and in addition the offset, caused by mismatch, in each of these $2^n - 1$ comparators causes the accuracy to drop down. Several improved flash architectures exist that reduce the number of building blocks, i.e. subranging, pipelined, folding and interpolating ADCs. Although they reduce the number of required building blocks and resistors, they do not fully solve the matching problematics and all still suffer from the same issues. For a clock frequency of 1 kHz the sample frequency in an organic flash ADC can reach 1 kHz but the estimated accuracy does not go above 3 bits.

4.2.2 Integrating ADC

Integrating ADCs are the other extreme in the trade-off between speed and accuracy. They are known to be very accurate but their sample rate is low. The working principle of a dual-slope integrating ADC is presented in Fig. 4.1b. The input signal is integrated on a capacitor C during the first clock phase ϕ_1. In the second phase

ϕ_2 the capacitor is discharged at a fixed speed, while a digital counter measures the discharge time. The stop signal is given by a comparator and the output of the counter is the n-bit digital word that represents the sampled analog input signal. A slightly modified quad-slope architecture exists which eliminates the influence of the offset in the comparator by measuring and later on subtracting the offset voltage from the actual sample. The price paid for this is that the quad-slope integrating ADC needs twice the time per sample compared to the dual-slope ADC.

The advantage of integrating ADCs is that they hardly suffer from mismatch. Moreover the offset present in the comparator can be eliminated in the quad-slope integrating architecture. As the same capacitor, the same integrator and the same comparator are constantly reused the integrating ADC performs, if so desired after calibration, with a high accuracy. The only disadvantage of this architecture is that it is known to be the slowest ADC architecture. Because of the limited headroom in terms of speed (1–10 kHz) in organic electronics technology, this architecture might not be the most preferred one.

The estimated accuracy of an organic integrating ADC reaches 8 bits. With this accuracy and on the assumption that the clock frequency is 1 kHz the sample frequency for the quad-slope architecture is only 1 Hz.

4.2.3 SAR ADC

The successive approximation register (SAR) ADC, presented in Fig. 4.2a is a pragmatic architecture that is built up with a DAC, a comparator and a digital processing unit, the SAR unit, that stores the digital word in a register. At the start of the conversion cycle the most significant bit (MSB) of the digital word is determined by comparing the sampled analog signal with the analog level that corresponds to the digital word $'100...0'$. The comparator output gives the correct value of the MSB, which is then set in the register. In the second clock cycle the $2nd$ MSB is determined through the same procedure. After n clock cycles the digital word is fully set and sent to the output of the ADC. This algorithm is a logarithmic approximation algorithm, i.e. after n clock cycles a 2^n precision is reached.

The n-bit DAC is typically built with a bank of resistors, a bank of current sources or a bank of capacitors. The mismatch in these banks limits the accuracy that can be reached with SAR ADCs, especially when resistors and current sources are employed. Capacitors in the organic electronics technology are known to have slightly better mismatch properties hence they are preferable. Whereas capacitive banks are prone to parasitic substrate capacitance in Si technology, these parasitic capacitors are not present on the non-conducting plastic substrate in the applied technology. The capacitive bank can be implemented through a C-2C ladder, which is beneficial for area. Nevertheless the mismatch properties of the capacitor bank still forms the weak link of this architecture. The estimated limit for the accuracy of the SAR architecture is around 6 bits. With a 1 kHz clock the SAR ADC sample rate is then limited to 167 Hz. A 6-bit C-2C integrated DAC is presented in Xiong et al.

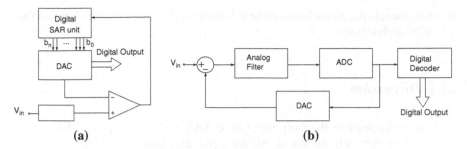

Fig. 4.2 Schematic system level diagrams of **a** the SAR ADC architecture and **b** the oversampling ADC architecture

(2010) and they applied it in a C-2C SAR ADC (Xiong et al. 2010) which was not fully integrated.

4.2.4 Oversampled ΔΣ ADC

The system level view of a ΔΣ ADC is presented in Fig. 4.2b. It is built up with an *nth*-order analog low-pass filter, an internal *b*-bit ADC and a *b*-bit DAC. Both of them can be any ADC or DAC architecture. After filtering a digital low-pass filter the highly oversampled output reaches a higher accuracy then the *b* bits of the internal ADC. The signal-to-noise ratio (SNR) of the ideal ΔΣ ADC is calculated from Eq. (4.1):

$$SNR_{\Delta\Sigma,\text{ideal}} = \frac{3\pi}{2} \cdot \left(2^b - 1\right)^2 \cdot (2n + 1) \cdot \left(\frac{\text{OSR}}{\pi}\right)^{2n+1} \qquad (4.1)$$

where *n* is the filter order, *b* is the number of bits in the internal ADC en DAC and OSR is the oversampling ratio in the ADC. The relation between OSR and SNR determines the speed-accuracy trade-off which is present in the ΔΣ ADC. All these three parameters improve the SNR when they are increased.

The ΔΣ ADC is even more insensitive to mismatch than the previous architecture since it simply does not employ a bank of matched components. Moreover due to the feedback loop the quantization error is not thrown away but it is subtracted from the input signal hence no information is lost. In this way, a ΔΣ ADC overcomes its own mistakes and adjusts internal imprecisions, e.g. offset in the comparator. By choosing the oversampling ratio of the ΔΣ ADC, the correct trade-off between speed and accuracy can be determined. The lowest circuit complexity of the converter is found when *n* and *b* are 1. With an OSR from 8 to 128 and a clock frequency of 1 kHz the signal bandwidth changes from 6.25 to 4 Hz and the SNR goes from 24 to 60 dB, which corresponds to 4-bit and 10-bit respectively. This formula, however, represents an ideal ΔΣ ADC with an infinite loop gain and the 10 bits accuracy is therefore unrealistic. Nevertheless these calculations illuminate the flexibility of the $\Delta\Sigma$ ADC architecture.

4.2.5 Discussion

In the previous sections the flash, integrating, SAR and oversampling ADC architectures were concisely presented. All the topologies have advantages as well as challenges that are intrinsically associated to the architecture. The flash ADC is probably the less interesting topology since it is only suitable for low precision and high speed and also because it employs a lot of hardware. Moreover the resistor bank and the parallel comparators are prone to mismatch.

The integrating ADC is a very slow architecture which can reach high precisions without suffering from mismatch. This is an interesting architecture for implementation in organic electronics technology, because the analog circuit complexity is limited.

The C-2C SAR architecture is a nice architecture in which a sensitivity to mismatch is present but under control in organic electronics technology. This topology mostly uses digital circuits. Therefore, it can be useful in organic electronic-technology.

The $\Delta\Sigma$ ADC has a very low mismatch sensitivity just like the integrating ADC. The feedback architecture automatically overcomes small variations without the need for tuning and trimming. Moreover this architecture has a very flexible trade-off between accuracy and speed through the oversampling ratio. For a low OSR, the difference between the speed of this architecture and the SAR architecture is limited. A comparison between the architectures in terms of their speed and accuracy is presented in Fig. 4.3.

Although both the SAR and the integrating ADC would also be good candidates for implementation, the $\Delta\Sigma$ ADC is the architecture that suits best within the subject of this research work. This suitable example of a more complex analog building block is further investigated in Sect. 4.3.

4.3 Design of a $\Delta\Sigma$ ADC

In Sect. 4.2 the $\Delta\Sigma$ ADC is introduced as the ADC architecture that is best suitable with the organic electronics technology for which it is intended. In this section the implementation of a first-order organic $\Delta\Sigma$ ADC and a second-order organic $\Delta\Sigma$ ADC with feedforward compensation is presented. Their system-level schematic diagram is presented in Fig. 4.4, where the order of the analog filter is 1 for the first-order ADC and 2 for the second-order ADC. The comparator and the level shifter are identical in both designs.

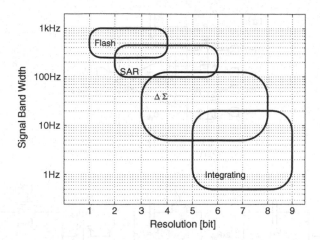

Fig. 4.3 Estimated speed-accuracy trade-off in four different ADC architectures in organic electronics technology with an assumed 1 kHz clock speed

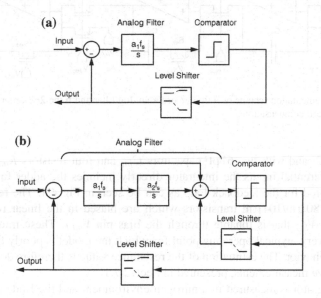

Fig. 4.4 Schematic system level implementation of **a** the 1*st*-order ΔΣ ADC and **b** the 2*nd*-order ΔΣ ADC with feedforward compensation

4.3.1 Analog First-Order Filter

The analog filter in a ΔΣ ADC is an integrator, i.e. a low-pass filter. The function of this integrator is to push the quantization noise, which is caused by the comparator, towards high frequencies. The differential implementation of the integrator is shown in Fig. 4.5a. It is built up with the differential 3-stage op amp, which is presented

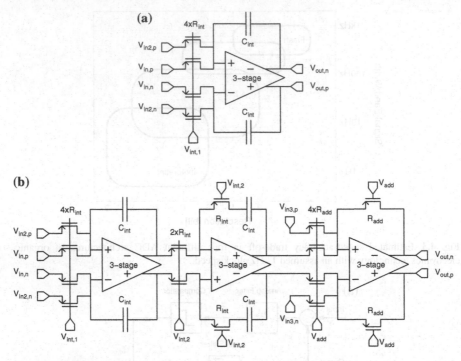

Fig. 4.5 Implementation scheme of **a** the 1*st*-order analog filter and **b** the 2*nd*-order analog filter with feedforward compensation

in Sect. 3.3.2, and with two 61 pF capacitors C_{int} and four resistors R_{int}. Through the two differential inputs the integrator directly includes the adder functionality that is required for the feedback loop as can be seen in Fig. 4.4. The resistors are built with 280 μm/104 μm transistors which are biased in the linear region with a very high V_{SG} that is tunable through the bias pin $V_{int,1}$. These transistors are biased in a very atypical operating point for which the model is poorly fitted to the transistor behavior. The estimation of the resistance value is therefore done directly from resistive measurements, presented in Sect. 2.7.1

The integrator is measured in a nitrogen environment and the Bode plot is presented in Fig. 4.6. The gain in the pass band is 17 dB which is 6 dB lower than the op amp gain. This is explained by the double differential input that acts like a resistive divider. The GBW of the integrator is 100 Hz and it consumes 21 μA from the 15 V power supply.

4.3.2 Analog Second-Order Filter

Similarly to the 1*st*-order filter an analog 2*nd*-order filter is designed for the 2*nd*-order ΔΣ ADC with feedforward compensation. As can be seen in Fig. 4.4

this filter requires a second integrator in which the feedforward path is integrated and an adder that brings in the second feedback path. For the implementation of this filter there are degrees of freedom available, i.e. the integration constants a_1 and a_2 and the feedback factor f. According to Schoofs (2007), the optimal values are 0.45, 0.5 and 0.32 for a_1, a_2 and f respectively. The implementation of the filter is presented in Fig. 4.5b that consists of three different op amp stages. The first stage is identical to the integrator in Sect. 4.3.1 and is not further discussed.

The second stage of the filter is an integrator with a parallel feedforward path, whose combined transfer function A_f is expressed in Eq. (4.2):

$$A_f = \frac{a_1 f_s}{s} + 1 \qquad (4.2)$$

where f_s is the sample frequency of the ADC that is estimated to be 1 kHz. This is implemented by an integrator with resistors placed in series with C_{int}, as shown in Sect. 4.3.1. The transfer function of this filter is calculated in Eq. (4.3):

$$A_f = \frac{R + \frac{1}{sC}}{R} = 1 + \frac{1}{sRC} \qquad (4.3)$$

which corresponds to the desired transfer function if the RC product is matched to $1/a_1 f_s$. This filter stage is built with 61 pF capacitors $C_{int,2}$ and with resistors $R_{int,2}$ that are also implemented with 280 μm/104 μm transistors biased in the linear region. Depending on the sample frequency the resistance of the integrators can be adapted through the bias pin $V_{int,2}$.

The third stage of this filter is an adder with two inputs for the forward and the feedback path. The aspired transfer function for both inputs is 1, which is reached with 280 μm/70 μm transistors biased in the linear region.

The measured transfer curves of the consecutive filter stages of the $2nd$ order filter are presented in Fig. 4.6. The expected behavior of the different filter stages is clearly visible. The transfer function of the first integrator also corresponds to the $1st$-order analog filter which is identical. The measurement of the adder is performed while both of its differential inputs are connected together. This explains why the measured gain of the adder, which is 3 dB, is above 0 dB. The bandwidth of the three filter stages is 1 kHz. Every stage of the filter consumes 21 μA from the 15 V power supply.

4.3.3 Comparator

The comparator is a critical block in the ΔΣ ADC. It must be able to detect very small differences of its input signal which is the output of the analog filter as can be seen in Fig. 4.4. The behavior of the comparator has an important influence on the ADC behavior and despite the fact it is a block with a digital output, it is here

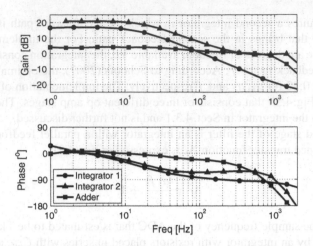

Fig. 4.6 Measured transfer curves of the three consecutive filter stages of the $2nd$-order filter. The transfer function of the first integrator also corresponds to the $1t$-order analog filter

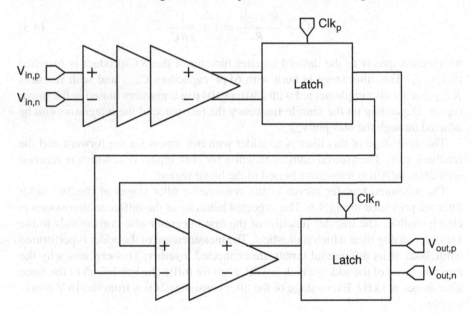

Fig. 4.7 Schematic view of the comparator that is employed in the $\Delta\Sigma$ ADC

considered from a fully analog point of view. The more gain the comparator has, the better the digitization of the filtered signal and the more accurate the digital output of the converter. As a consequence the lower boundary for the gain in the comparator is determined by the desired accuracy of the ADC. This boundary is given by the presupposed signal-to-noise ratio (SNR) of the ADC. Since the applied organic electronics technologies are unipolar it is not obvious to make a comparator

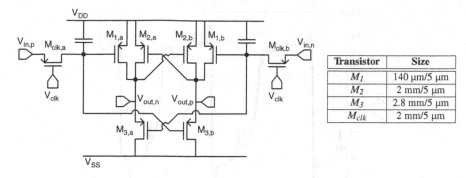

Fig. 4.8 Implementation of the differential latch and a list of the transistor sizes

Transistor	Size
M_1	140 µm/5 µm
M_2	2 mm/5 µm
M_3	2.8 mm/5 µm
M_{clk}	2 mm/5 µm

reliable and with high gain at the same time and special attention must go to offset compensation. The comparator employed in the $\Delta\Sigma$ ADC is shown in Fig. 4.7. It is built with the single-stage amplifier, presented in Sect. 3.3.1, which is specifically designed for high gain and reliability. To ensure a good behavior 3 consecutive stages of this amplifier, which are sized up with a factor 3, are employed. The high-pass filters between the consecutive stages enable the suppression of the DC offset that is present in each stage. Consequently a latch is introduced to synchronize the comparator signal to the clock signal. Finally this latch is followed by a chain of 2 more amplifiers that ensures a full-swing signal and a second latch that is clocked with the inverted clock signal.

The implementation of the latch is presented in Fig. 4.8. It is built with two cross-coupled inverters, M_2 and M_3, and two pull-up transistors M_1 that switch the latch. Furthermore two pass gates M_{clk} are present that are driven by V_{clk} and two capacitors C_{in} that store the sampled input signal.

The measured output of the latch is shown in Fig. 4.9. The measurement is performed at a 1 kHz clock speed with a 500 kHz input signal. It is clearly visible that the limiting factor for the speed is the pull-down behavior of the latch.

4.3.4 Level Shifter

The level shifter is the last building block employed in the presented $\Delta\Sigma$ ADC. Its function is threefold:

In the first place this building block shifts the mean DC value of the digital feedback signals to the correct DC value which is desired at the input of the analog filter.

In the second place the two digital levels determine the digital swing of the feedback signals, which also influences the voltage range in which the analog input signal may be situated. Moreover, in order to make this test design tunable it is very useful to be able to change these digital levels externally.

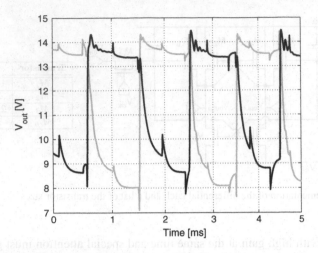

Fig. 4.9 Measured differential output of the presented latch

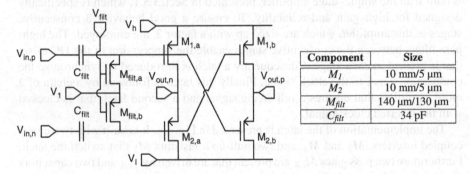

Component	Size
M_1	10 mm/5 μm
M_2	10 mm/5 μm
M_{filt}	140 μm/130 μm
C_{filt}	34 pF

Fig. 4.10 Implementation of the level shifter and a list of the transistor sizes

Finally, in the third place, this level shifter has a buffer functionality that can drive the TL084 op amp which is connected in unity-feedback configuration and reads out the chip output on the test setup.

The implementation of the level shifter is shown in Fig. 4.10. The two inverters, which are differentially driven, make the output signals switch between the desired digital levels, V_l and V_h. Two high-pass filters ensure that the switching signals are brought to the appropriate level V_1. Since the input swing is larger than the output swing the level shifter improves the digital character of the output signals. This circuit has not been separately measured because its complexity is low and its behavior straightforward.

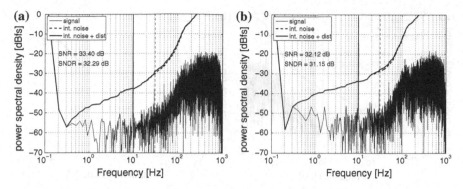

Fig. 4.11 Simulated output power spectral density of **a** the presented first-order $\Delta\Sigma$ ADC and **b** the presented second-order $\Delta\Sigma$ ADC

4.3.5 Simulation Results

The simulation results of both the first- and second-order $\Delta\Sigma$ ADC are given in Fig. 4.11. The simulated SNR of the $1st$-order ADC is 33 dB. The simulation is performed with a 1 kHz clock frequency and a sinusoidal 10 Hz input. The SNR is calculated for an OSR of 16. The simulated SNR of the $2nd$-order ADC is slightly lower than the SNR of the $1st$-order ADC, whereas this is expected to be higher. This is explained by the operating points which are not ideal and by the limited gain of the analog filter and the comparator which determines the noise floor of the PSD. Furthermore, the simulated noise shaping slope of the $2nd$-order ADC is not steeper than that of the $1st$-order ADC. This can also be explained by the operating points which are not optimal.

The added value of these simulations is limited, since the models are poorly fitted. The most important conclusion that can be taken from these simulations is that the circuits perform in a conceptual way. The circuits are made in a tunable way so that their behavior can be further optimized during the measurements.

4.3.6 Measurement Results

The measurements of the $\Delta\Sigma$ ADC have been performed in an N_2 environment at room temperature. Figure 4.12a shows the PSD of the measured output of the $1st$-order $\Delta\Sigma$ ADC for a $2\,V_{ptp}$ 10 Hz input sine wave and a 500 Hz clock frequency. An SNR of 26.5 dB is measured with an OSR of 16. The losses of precision against simulation under the same conditions are caused by limited open-loop gain in the integrator and the comparator and low accuracy of the applied models in the simulations. The input BW for an OSR of 16 is 15.6 Hz. The limitation towards low frequencies is determined by the high-pass filters in the circuit and lies around 0.1 Hz. In Fig. 4.12b

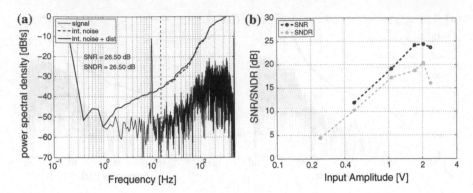

Fig. 4.12 **a** Measured output power spectral density of the presented 1*st*-order ΔΣ ADC with a 2 V$_{rmptp}$ 10 Hz sinusoidal input and a 500 Hz clock frequency. **b** The measured SNR and SNDR for a 3 Hz input with a varying amplitude at a 500 Hz clock frequency

the measured SNR and SNDR of the ADC are visualized versus the input amplitude of a 3 Hz input signal. In this figure the linear relation between input amplitude and SNR/SNDR is clearly visible. The peak SNR for 3 Hz input is 25 dB. The 1 dB loss, compared to the SNR for a 10 Hz input, is probably caused by the high-pass filters in the circuit. A difference between SNR and SNDR of about 3 dB is present, even for low amplitudes. This might be caused by distortion in the analog filter or in the feedback circuit. The circuit consumes 100 μm from a 15 V power supply. Table 4.1 summarizes the measurement results. The chip photograph of the presented circuit is shown in Fig. 4.13.

Due to the high-pass filters in the single-stage amplifiers and in the throughput between consecutive stages there is not enough phase margin for low frequencies. This means that the continuous-time behavior of this circuit is unstable at frequencies around 0.1-1 Hz. A precharge-and-hold functionality, as discussed in Sect. 3.3.2.1, is therefore applied during these measurements, which are then performed in the hold phase. The 2*nd*-order ADC is even more unstable for low frequencies due to the higher number of low-frequency poles in the 2*nd*-order analog filter. As a consequence the frequency of the oscillation becomes higher and the hold phase in the precharge-and-hold measurement setup is not long enough to carry out the measurements.

4.4 Discussion

As outlined in Sect. 4.3.6 the high-pass filters in the presented designs cause unstable behavior at low frequencies. In the 1*st*-order ADC the problem was overcome through a precharge-and-hold way of measuring and good measurements have been performed whereas in the 2*nd*-order architecture even with this technique it has

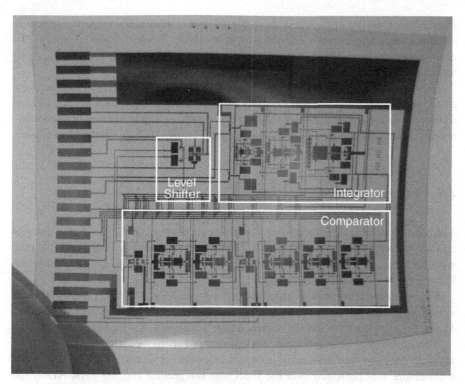

Fig. 4.13 Chip photograph of the presented 1*st*-order $\Delta\Sigma$ ADC on foil. The different building blocks of the ADC are clearly distinguished. The chip area is $13 \times 20\,mm^2$. This circuit is made in technology provided by Polymer Vision.

Table 4.1 Summary of the measurement results for the 1*st*-order $\Delta\Sigma$ ADC on foil (Marien et al. 2010, 2011)	Specification	Value
	Power Supply	15 V
	Current Consumption	100 μA
	Clock Frequency	500 Hz
	Input Frequency	10 Hz
	SNR@ 10 Hz	26.5 dB
	SNDR@ 10 Hz	24.5 dB
	OSR	16
	Band Width	15.6 Hz
	Chip area	$13 \times 20\,mm^2$

turned out impossible to obtain qualitative measurements since the hold time was not long enough to perform the measurements. This problem is inherently connected to the use of high-pass filters in a signal path and can only be solved by omitting these filters. One solution for this problem is provided in Sect. 2.4.3.1 and is employed and demonstrated in an operational amplifier in Sect. 3.3.3. By employing the functionality of the backgate of the 4-contact transistors a direct connection of consecutive amplifiers in a signal path is enabled. This technique is expected to also further improve the precision of organic ADCs, especially the $\Delta\Sigma$ ADCs, since higher-order filters can be designed in a reliable way with it.

Furthermore it is worth mentioning that all the high-pass filters have been employed to overcome one and the same problem during the design, i.e. the lack of a complementary set of transistors. As soon as reliable complementary organic technology arises, the complexity of analog organic circuits will significantly increase.

4.5 Conclusion

The concept of organic smart sensor systems requires both analog an digital circuits and for the mutual transition ADCs and DACs are required. The goal in this chapter was the fabrication and the demonstration of organic ADCs by using the gathered knowledge about amplifier design in Chap. 3 .

First an introduction was provided about the application field of organic ADCs and system requirements were formulated as a guide line during the design procedure. Most of the applications did not require a bandwidth which is higher than 10 Hz.

Subsequently a comparative study was presented about the different existing ADC architectures and their feasibility in organic electronics technology. The $\Delta\Sigma$ ADC architecture was preferred over the flash, the integrating and the SAR ADCs for reasons of variability and for the flexible trade-off between speed and accuracy.

Then the implementations of a $1st$-order and a $2nd$-order $\Delta\Sigma$ ADC on foil were elucidated and their measurement results were presented. The precision of the $1st$-order ADC amounted to 26.5 dB and a signal bandwidth of 15.6 Hz was obtained with a clock frequency of 500 Hz.

Finally a discussion was given about the quality of the presented designs. The high number of high-pass filters in the designs was explained to have an unfavourable influence on the stability at low frequencies of the circuits. This problem was investigated and further improvements were supplied.

References

Marien H, Steyaert M, van Veenendaal E, Heremans P (2011) ADC design in organic Thin-Film electronics technology on plastic foil in International Workshop on ADC (IWADC). Orvieto, Italy, June

Xiong W, Guo Y, Zschieschang U, Klauk H, Murmann B (2010) A 3-V, 6-Bit C-2C Digital-to-analog converter using complementary organic thin-film transistors on glass. IEEE J Solid-State Circuits 45(7):1380–1388

Xiong W, Zschieschang U, Klauk H, Murmann B (2010) A 3V 6b successive-approximation ADC using complementary organic thin-film transistors on glass. IEEE international solid-state circuits conference digest of technical papers (ISSCC), pp 134–135.

Schoofs R (2007) Design of high-speed continuous-time delta-sigma A/D converters for broadband communication. PhD Thesis, ESAT-MICAS, K.U.Leuven, Belgium.

Marien H, Steyaert M, van Aerle N, Heremans P (2010) An analog organic first-order CT $\Delta\Sigma$ ADC on a flexible plastic substrate with 26.5dB precision. IEEE international in solid-state circuits conference digest of technical papers (ISSCC), pp 136–137.

Marien H, Steyaert MSJ, van Veenendaal E, Heremans P (2011) A fully integrated $\Delta\Sigma$ ADC in organic thin-film Ltransistor technology on flexible plastic foil. IEEE J Solid-State Circuits 46(1):276–284

Chapter 5
Sensors

The properties of organic electronics technology are very well suited with the demands of flexible display and smart sensor system applications. However, this technology is nowadays still sensitive to variations of the behavioral parameters. This sensitivity of organic semiconductors to environmental changes is predominantly experienced as adverse but it can also beneficially be employed in sensors. For smart sensor systems it is more than clear that sensors are required but sensors, i.e. touch sensors, can also in an indisputable way make flexible display devices more user-friendly by adding user control, e.g. through touch pads.

In this chapter the implementation of sensors in organic electronics technology is discussed. This corresponds with the first building block of block diagram of the proposed organic smart sensor system in Fig. 1.5. In Sect. 5.1 the application field of sensors is explored. An overview of the state-of-the-art is given in Sect. 5.2 concerning sensors implemented in organic electronics technology. Section 5.3 introduces the existing touch pads architectures. Subsequently in Sect. 5.4 the designs of a one-dimensional (1D) and a two-dimensional (2D) capacitive touch sensor on foil are presented and measurement results of both circuits are presented. Section 5.5 discusses those results and the performance of the sensors. Finally the conclusions of this chapter are drawn in Sect. 5.6.

5.1 Application Field

The expected application field of organic sensors is guided by the main properties of organic electronics technology. The flexibility of the plastic substrate in the first place enables the use of sensors in applications where flat and rigid Si substrates can not be used, e.g. on bent or flexible surfaces like the human skin. The expected low cost per area of the plastic substrates, in the second place, enables the large-area sensor applications, e.g. artificial skin (Kawaguchi et al. 2005) which is applicable in robots. One of the expected areas of interest, which is directly induced by the technological

H. Marien et al., *Analog Organic Electronics*, Analog Circuits and Signal Processing,
DOI: 10.1007/978-1-4614-3421-4_5, © Springer Science+Business Media New York 2013

features is health monitoring. Due to the proportional rise of the ageing population in our society there is a need for health monitoring devices that lower the cost of continuous medical supervision, e.g. to detect when a patient falls. Other future biomedical applications are found in the monitoring of heart beat, blood pressure, wound healing and others. It is an advantage in this domain that most of the signals of interest in the human body have a low bandwidth which can already be dealt with by the existing organic technologies at this moment.

Besides the health monitoring branch of future applications there is also an interest for monitoring the quality of food during its life time, i.e. before it is consumed. This life time is characterized by a set of processing and packaging operations, by transportation and by storage on different locations, e.g. in the supermarket and at home in the fridge. With the use of adequate sensor systems the quality of the products can be monitored, ensuring to the consumer that a quality product is delivered, e.g. that the product has been sufficiently cooled along the road or that the milk has not yet turned sour.

Another interesting branch for sensors is found with touch sensors in flexible display devices. These touch screens increase the user experience and have become indispensable for handheld consumer electronics such as smart phones and tablet PCs. Besides all the mentioned applications in this section there is an even larger domain of interest for organic sensors that is left for the reader's creativity.

5.2 State-Of-The-Art

Organic electronics technology is known to be sensitive to a broad range of external influences. The sensitivities of pentacene, which is the semiconductor employed in this dissertation, have been discussed in Sect. 2.5. These sensitivities are mainly experienced as a drawback since the behavior of standard circuits suffers from the change of transistor parameters. Yet it opens gates for the design of organic sensors. This interesting domain has been investigated before. By examining and isolating a certain sensitivity and creating specific layers with special properties, organic sensors have been made. In this section an overview is given of the state-of-the-art in organic sensor applications.

5.2.1 Temperature Sensors

An integrated 3×3 array of temperature sensors has been reported in He et al. (2010) employing p-type transistors only. The sensor relies on the current of an organic transistor biased in the subthreshold region. One observes two important differences between the current in an OTFT and in a MOSFET.

- The current in an OTFT increases with temperature in both subthreshold and above threshold regimes, whereas in MOSFETs the above-threshold current decreases with increasing temperature. As a result, a linear dependency of the temperature with a negative slope is obtained when measuring a differential voltage.
- The responsivity of the sensor is about 20 times higher in OTFTs than in MOSFETs.

A second temperature sensor has been mentioned in Kawaguchi et al. (2005). In that reference measurement results are presented but the temperature sensor design has not been discussed.

5.2.2 Chemical Sensors

Chemical sensors built with pentacene OTFTs have been broadly reported in literature and are able to sense a variety of molecules. A pentacene transistor employed as a sensor for ammonia (NH_3) gas has been reported in Zan et al. (2011). The sensor is designed for the analysis of breath samples for the diagnosis of liver cirrhosis and renal failure. Through measurements the selectivity of the sensor to ammonia, neglecting other gases, such as alcohol, carbon dioxide and acetone has been demonstrated. Similar sensors have been presented in Mori et al. (2009); Liu et al. (2009) where alcohol is the chemical of interest. According to Mori et al. (2009); Subramanian et al. (2006) the selectivity of chemical sensors can be changed by chemically modifying the gate dielectric.

5.2.3 Pressure Sensors

Several techniques have been published for making pressure sensors in pentacene-based technology. In Darlinski (2005) the pressure dependence of pentacene transistors on a glass substrate is investigated. A transistor with a $10\,\mu m$ channel length is measured without and with a mechanical pressure of 10^6 pa applied on top of the transistor. The current in the transistor which is held under pressure is significantly higher than the pristine transistor when V_{GS} is 0 V. The presented changes take seconds hence the behavior is attributed to trap states in the semiconductor/dielectric interface which change the threshold voltage, the mobility and the contact resistances of the transistor. These results have been confirmed by Manunza et al. (2006) which reports a similar technique with pentacene transistors implemented on a flexible substrate.

Another technique has been reported by Kawaguchi et al. (2005) which extends the technology with a rubbery material with a pressure dependent resistivity. A printed 16×16 pixel active matrix is employed to read out the resistivity at each pixel. The sample time of a pixel is 30 ms which means that the full 16×16 pixel array is measured after about 2 s. This pressure sensor has been designed for large-area applications, such as artificial skin for robots. A photograph of the sensor placed on top of a robotic hand is shown in Fig. 5.1.

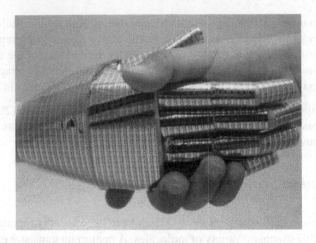

Fig. 5.1 Photograph of the large-area pressure sensor matrix placed on a robotic hand as artificial skin, reported by University of Tokyo (Kawaguchi et al. 2005; http://www.ntech.t.u-tokyo.ac.jp. 2012)

5.2.4 Other Sensors

Several other kinds of sensors have also been reported. Their functionality is always based on the sensitivity of the transistor characteristics which are influenced by external parameters. A vertically stacked photo-sensor with tunable optical properties has been presented in Jeong et al. (2010). Furthermore moisture sensors (Lo & Tai 2007) and pH sensors (Caboni et al. 2009) have been reported.

Most of the existing sensors have been implemented using the properties of the semiconductor, i.e. pentacene, or by the enhancement of a technology with one or a set of additional special-purpose layers. Of course it is also possible to make capacitive sensors in an existing technology. Such tactile sensors measure the proximity of a conductive material and are very useful for sensing a position, e.g. of a finger. This technique does not rely directly on the properties of the organic semiconductor hence capacitive sensors are insensitive to parameter variations. This technique has been implemented in this work and measurement results are presented in Sect. 5.4. A similar technique has very recently been applied in a capacitive touch pad on a glass substrate in Kim et al. (2009).

5.3 Touch Pad Architectures

Touch pads are broadly used in today's consumer devices for user control. Nowadays touch screens are getting even more distributed for an even better user experience. With the launch of flexible displays in sight, in this work a flexible touch pad is considered. Both a flexible $1D$ linear touch pad and a flexible $2D$ touch pad are designed and measured. As an overall introduction to flexible touch pads in this

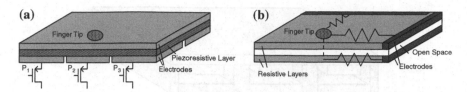

Fig. 5.2 **a** Working principle of a piezoresistive touch pad. The resistance of the pixel changes when a finger is placed on top. Every pixel is separately measured through an active matrix. **b** Working principle of a resistive touch pad with two parallel resistor layers. The distance between the finger and the electrode determines the resistance which is measured. This technique can be applied in the two dimensions

section an overview is given of the most often used touch pad architectures, i.e. resistive touch pads and capacitive touch pads.

5.3.1 Resistive Touch Pad

A resistive touch pad employs a resistive effect to locate the position of the finger. One possible way to implement such a resistive touch pad is to use a piezoresistive layer, i.e. of which the resistance is pressure sensitive, like in Kawaguchi et al. (2005). This working principle of this technique is shown in Fig. 5.2a, where every pixel measures the resistance of the resistive layer on that position. The read-out circuit is represented by the transistor and not further discussed. An alternative way to make a resistive touch pad is with a resistive layer. When a finger touches this touch pad this layer is pushed against a parallel resistive layer (Bai & Chen 2007). The working principle of this technique is visualized in Fig. 5.2b. Here the position of the finger is calculated from the measured resistance which is proportional to the distance between the electrodes and the finger. This technique can be repeatedly applied in two orthogonal dimensions in order to obtain two-dimensional information of the location of the finger. Both of the mentioned techniques require the use of additional special-purpose layers and are not easily implemented in the existing organic electronics technologies.

5.3.2 Capacitive Touch Pad

Capacitive touch pads detect the change of the capacitance of the touch pad. The latter is caused by a finger which is placed on top of the touch pad (Kim et al. 2009; Wang & Ker 2011). By measuring the capacitance on every pixel in the pixel array of the touch pad the position of the finger is extracted. The capacitance can be mechanically changed by pushing two plates closer to eachother. The working

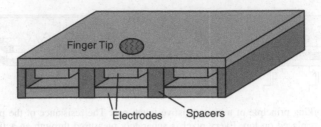

Fig. 5.3 Working principle of the capacitive touch pad with a mechanical change of the capacitance. When a finger pushes on a pixel the distance between the capacitor plates of that pixel decreases and the capacitance increases

principle of this technique is shown in Fig. 5.3. The thickness of the elastic material decreases when a finger is placed on top of a pixel hence the capacitance of that pixel increases.

Another technique which is often used, employs two adjacent metal planes with a very low mutual capacitance. When a finger is positioned on top a series connection of two capacitors is created. This capacitance can be measured through several possible techniques. The touch screen of an iPhone (Wilson 2007) is based on this capacitive principle. The capacitive technique is also employed for the flexible touch pad design in this work and further discussed in Sect. 5.4.

Recently, work on a flexible organic 2D touch pad has been published (Yokota et al. 2011). There multiple flexible foils with different functionalities are put on top of each other. The top layer is a pressure-dependent material that generates a voltage drop of ~10 mV. The next layers are an array of capacitors, an array of pseudo-CMOS inverters and an active-matrix of transistors respectively. The working principle of this sensor is demonstrated. It senses the 10 mV voltage drop, caused by the touch of a finger up to a 150 mV output voltage.

5.4 Designs

The field of flexible sensors in organic electronics technology has been explored in this work through 1D and 2D capacitive touch sensors. Contrarily to the largest part of the state-of-the-art in Sect. 5.2 where technologies are extended with additional layers with specific properties, the goal in this work is to integrate sensors in an existing organic electronics technology without the addition of special-purpose layers. This is done through the design of capacitive touch sensors in a technology which is optimized for active matrix rollable displays (Gelinck et al. 2004; Gelinck 2005). Such sensors can increase the user-friendliness of such devices as they fit seamlessly in the given technology.

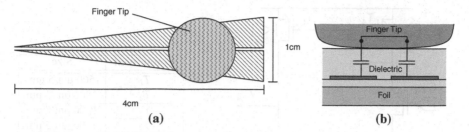

Fig. 5.4 **a** Top view of the capacitive sensor. **b** Cross-sectional view of the series capacitors created when a finger is located on top of the sensor

Table 5.1 Implementation of the sensor read-out circuit and a list of the component sizes

Component	Size
T_1	$500\,\mu m/5\,\mu m$
T_2	$1000\,\mu m/5\,\mu m$
T_3	$1000\,\mu m/5\,\mu m$
T_4	$1000\,\mu m/5\,\mu m$
T_5	$250\,\mu m/5\,\mu m$
C_{out}	$25\,pF$

5.4.1 1-Dimensional Touch Sensor

5.4.1.1 Architecture

Figure 5.4 shows the architecture of the 1D capacitive touch sensor. It is built with two triangular metal plates which are both implemented in the second metal layer Met_2. Each capacitor plate is a right triangle with a width of 4 cm and a 0.5 cm height. In the untouched set-up only a very small hence negligible capacitance is experienced between these two plates. When a finger is present on top of these plates, a series connection of two capacitors is obtained through the finger. The capacitance per area between Met_1 and Met_2, and between Met_2 and Met_3 is given by Gelinck (2005) and amounts to $10\,nF/cm^2$ and $2\,nF/cm^2$ respectively. The size of a finger tip is estimated as a circle with a 1 cm diameter. When it is positioned at the wide part of the sensor and supposed to be fully conductive an estimated series connection of two overlap capacitors with a net capacitance C_{sens} of ~500 pF is experienced. When the finger is at the smallest side of the sensor, C_{sens} decreases to 0 pF. In reality the maximal value of C_{sens} is lower since dry skin is rather an insulator than an ideal conductor which increases the distance between the capacitor plates and consequently reduces the C_{sens}. In this way the capacitance between the metal plates is a measure for the location of the finger.

The sensor read-out circuit is presented in Fig. 5.5. It is built with three switches $T_{1,2,3}$, a current mirror $T_{4,5}$, a variable capacitor C_{sens} and a fixed 25 pF output capacitor C_{out}. The working principle of this architecture consists of two phases. In the reset phase, Clk_2 is low and Clk_1 is high, hence transistors T_1 and T_2 are active

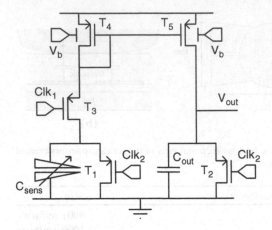

Component	Size
T_1	500 μm/5 μm
T_2	1000 μm/5 μm
T_3	1000 μm/5 μm
T_4	1000 μm/5 μm
T_5	250 μm/5 μm
C_{out}	25 pF

Fig. 5.5 Implementation of the sensor read-out circuit and a list of the component sizes

and the capacitors C_{sens} and C_{out} are discharged. In the sample phase Clk_1 is low and Clk_2 is high, hence T_3 is switched on and enables the charging of the capacitor C_{sens}. The charging current is subsequently copied through the current mirror with a tunable current, through V_b, and integrated on the output capacitor. The output voltage at the end of the sample phase is then a measure for the capacitance value of C_{sens} hence a measure for the location of the finger. It is determined by the ratio of the capacitors C_{sens} and C_{out} and by the ratio of the current mirror $T_{4,5}$ and is given by Eq. (5.1):

$$V_{out} = B \cdot \frac{C_{sens}}{C_{out}} \cdot V_{DD} \quad \left(with \ B = \frac{\frac{W_5}{L_5}}{\frac{W_4}{L_4}} \right) \tag{5.1}$$

Since the output voltage is only a function of the ratio of capacitors it is very insensitive to the degradation effects which affect the V_T.

The capacitance of C_{sens} is in a linear way proportional to the location of the finger, as can be seen in Fig. 5.6 where the overlap area of the finger, implemented as a circular disk, is simulated as a function of its location. Furthermore in this figure the simulated output voltage is shown as a function of the finger location. A linear fit for this sensor behavior has been extracted and included in the figure. The difference on the X-axis between the simulated output curve and the linear fit for a certain output voltage is a measure for the precision of this sensor. The largest error is obtained for an overlap area of 45 mm^2 on the Y-axis and amounts to 2.5 mm. As a result, the simulated error of this sensor is always smaller than 2.5 mm.

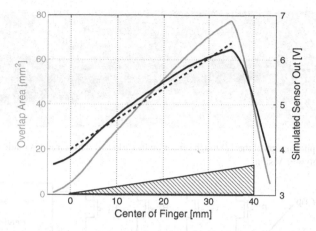

Fig. 5.6 Simulated overlap area (*gray*) when an ideal finger is positioned on top of the sensor and the simulated output voltage (*black*) of the sensor read-out. The *dashed line* corresponds with a linear fit of the sensor output. The largest error is obtained for an overlap area of 45 mm² on the *Y*-axis and amounts to 2.5 mm

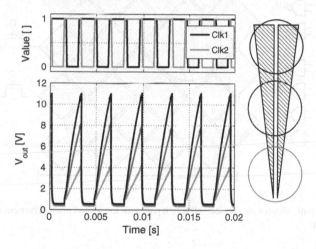

Fig. 5.7 Measured output of the 1*D* sensor for three different positions of the finger on the sensor

5.4.1.2 Measurement Results

The mesurement results of the 1*D* touch sensor are presented in Fig. 5.7. The three output signals correspond to the finger positioned at the small side, in the center and at the widest side of the sensor. It is clearly visible that the sensor output is higher when more overlap capacitance is created. The sample rate of this measurement is 330 S/s. The measurement is performed at a supply voltage of 15 V and the sensor read-out draws 6 μA, hence the sensor consumes 90 μW. The measurement results are summarized in Table 5.2. A chip photograph of the sensor is shown in Fig. 5.8.

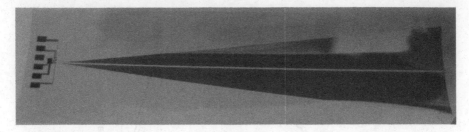

Fig. 5.8 Chip photograph of the $1D$ touch sensor on flexible foil. This circuit is made in technology provided by Polymer Vision

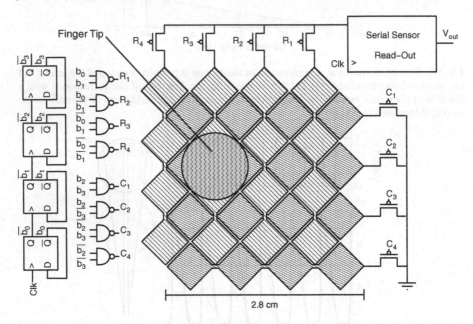

Fig. 5.9 Schematic view of the flexible $2D$ touch pad with integrated pixel driver circuit and sensor read-out circuit

5.4.2 2-Dimensional Touch Sensor

5.4.2.1 Architecture

The same technique of capacitive sensing, which is used in the $1D$ sensor can also be extrapolated towards a 2-dimensional touch sensor, where the read-out circuit measures the capacitance of the different pixels in a serial way. This is demonstrated with the design of a flexible $2D$ touch pad with 4×4 pixels presented in Fig. 5.9. The row and column plates are respectively built in the first and the second metal layer and they are built with $4.9 \times 4.9\,\text{mm}^2$ squares. When a finger is positioned on top of a pixel, as depicted in the figure, an overlap capacitance is generated according

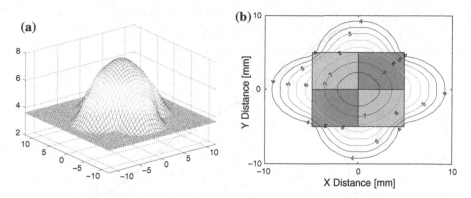

Fig. 5.10 a 3D plot of the simulated sensor output of a pixel as a function of the position of a finger. **b** Contourplot of the simulated sensor output

to the same principle, which was explained in Fig. 5.4b. The sensor read-out circuit is the same circuit as in Fig. 5.5, but now a row and column selector switches R_{1-4} and C_{1-4} are included which activate all the pixels one-by-one for a serial evaluation of each position's capacitance. Each row switch is active for 1 clock cycle and closed for the 3 subsequent clock cycles whereas each column switch is active for 4 subsequent clock cycles and closed for the 12 next clock cycles. The selector switches are driven with active low signals generated by a selector circuit, built up with a binary counter and eight NAND gates.

The sensor read-out is an analog building block and the output contains more information than just the digital signal. The analog output is a measure for the relative position of the finger, i.e. when the overlap is only partial. As a result the precision of the sensor can be increased through interpolation. The accuracy of interpolation depends on the linearity of the sensor read-out. Simulations have been performed to estimate this linearity. The simulated output voltage of a pixel measurement is visualized in Fig. 5.10. Due to the square form of the pixel it is logical that the steepness of the function varies with the direction. In Fig. 5.11 the simulated differential output is described on the direct path between two adjacent pixels. This curve is obtained by subtracting the output signals of both pixels. A linear fit of this curve is included in the figure. The largest positional error caused by the nonlinearity of the sensor amounts to 0.3 mm. It is encountered at a value of ±1.2 V on the Y-axis. This result gives an idea of which precision a digital processing unit can extract from the sensor through linear interpolation techniques.

5.4.2.2 Measurement Results

In Fig. 5.12 the measurement results of the $2D$ touch pad are presented. Figure 5.12a shows the output signal for two different positions of the finger, i.e. $R_1 C_1$ and $R_3 C_3$,

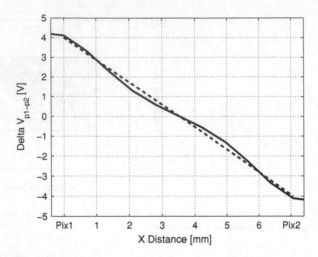

Fig. 5.11 Simulated differential sensor output for two adjacent pixels

referred to the reset signal of the counter. The measurement of position R_1C_1 corresponds with the first peak after the clock reset as can be seen. Accordingly the pixel on position R_3C_3 is measured at the 11th clock cycle. The measurement has been performed at a sample frequency of 150 S/s. The measured cut-off frequency of the sensor is situated around 1.5 kHz. At this sample rate and for a 4 × 4 pixel sensor the frame rate is 93 Hz which is much higher than the bandwidth of the motion of a finger which amounts to a few hertz only. With a minimal frame rate of 20 Hz the extrapolation of this sensor towards 8 × 8 pixels would still make sense. This is, however, not further investigated.

In Sect. 5.4.2.1 simulation results are presented which demonstrate the further digital interpolation of the output signals. This increases the precision of the sensor. This interpolation is confirmed by the measurements shown in Fig. 5.12b that are performed with a finger halfway in between positions R_3C_2 and R_3C_3 and with a finger right in the middle of the pixels R_1C_1, R_1C_2, R_2C_1 and R_2C_2.

The selector circuit has been measured separately and the results are presented in Fig. 5.13 where all the driving signals for the switches R_{1-4} and C_{1-4} are depicted. Each row signal is in turn low for one clock cycle and then high for the three subsequent clock cycles. The same holds for the column signals yet 4 times slower as can be seen in the figure. It is remarkable that the expected high signals are often situated only halfway the supply voltage. This is due to the NAND gates and it can easily be solved in a redesign with additional buffer stages for signal recovery.

All the measurements of the $2D$ sensor have been performed in ambient environment at room temperature and at a power supply voltage of 15 V. They are summarized in Table 5.2. The chip photograph is shown in Fig. 5.14.

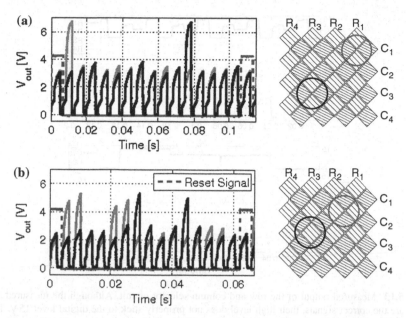

Fig. 5.12 a The measured output of the $2D$ touch pad with the finger on two different positions $R_1 C_1$ and $R_3 C_3$. The measurement is performed with a power supply voltage of 15 V and at a sample rate of 150 S/S. **b** The measured sensor outputs when the finger is halfway between positions $R_3 C_2$ and $R_3 C_3$ and in the center of $R_1 R_2 C_1 C_2$ at a 270 S/s. Both measurements have been performed in ambient with external selector signals and with a shielded finger

5.5 Discussion

The sensors presented in Sect. 5.4 are a proof of concept for electronic sensors implemented in an organic electronics technology. They prove that not only at the technology level but also on the circuit level sensor design is possible in a given organic technology without changing a single parameter in the production flow. However, a few improvements of this sensor should be included in a redesign to obtain full and correct operation. In the first place the selector circuit is not fully functional because its output signals are not nicely digital. In some cases, their output voltage is around $\frac{V_{DD}}{2}$ whereas it is expected to be high, i.e. equal to V_{DD}. This is caused by $NAND$ gates that are not powerful enough to pull up the output signals when one of their inputs is low and the other high. This problem can easily be solved in a redesign through proper buffering of the digital signals with inverters.

In the second place a problem arises when the $2D$ sensor is touched with a human finger. This is caused by the large capacitance (typically 100 pF) of the human body towards the environment. This capacitor creates a parasitic parallel path that hampers the proper read-out of the different column lines. As a result, four peaks appear in the serial output, instead of one expected peak. Those four peaks correspond to the four pixels in the same row as the position of the finger. Practically, the circuit can

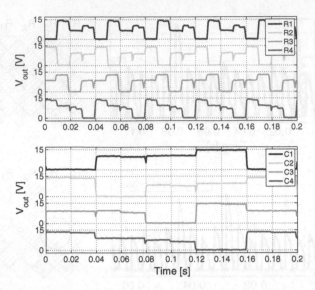

Fig. 5.13 Measured output of the row and column selector circuit. Although the measured signals are the correct signals, their high level does not properly stick to the digital level 15 V. This measurement is performed at a clock frequency of 100 Hz

Fig. 5.14 Chip photograph of the 2*D* touch sensor on flexible foil. This circuit is made in technology provided by Polymer Vision

only differentiate between the rows and the dimension of the sensor is reduced from 2 to 1. This is schematically shown in Fig. 5.15 where the parasitic path through the finger is present, even when switch C_i is off, hence every pixel located on row R_j is sensed as active. Solutions for this problem do exist, e.g. the use of synchronous

Table 5.2 Summary of the measurement results for both the $1D$ and the $2D$ touch sensor integrated on plastic foil (Marien et al. 2012)

Specification	1D Sensor	2D Sensor
Power Supply	15 V	
Current Consumption	6 μA	
Sample rate	1.5 kS/s	
Frame rate	–	93 Hz
Simulated Precision	2 mm	7 mm
After Interpolation	–	0.3 mm
Power	90 μW	
Chip Area	$1.0 \times 4.4\, cm^2$	$3.5 \times 3.5\, cm^2$

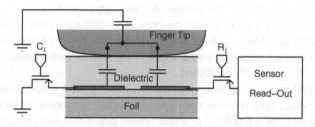

Fig. 5.15 Schematic view of the parasitic path through the finger that disturbs the $2D$ sensor

detection to measure the capacitance. This technique increases the complexity of the sensor and a redesign is required. A more ad rem solution for this problem which has been employed during the measurements is to shield the finger by using an isolating finger cap with a conductive layer on top. With this technique the parasitic path is cut and the $2D$ sensor performs correctly.

A last point of discussion is that the sensor area is not transparent. This is due to the gold layer that is used for the metal layers. Transparent materials, e.g. indium-tin oxide (ITO) (Lapinski et al. 2009) or Poly(3,4-ethylenedioxythiophene) poly(styrenesulfonate) (PEDOT:PSS) (Halik et al. 2002), exist which are conductive and integratable in similar technologies. As explained previously in this section it is not the goal of the presented sensor designs to change the production process but rather to use the existing technology and demonstrate the feasibility of sensors in a fixed organic electronics technology.

5.6 Conclusion

In our society with a graying and increasing population the future availability of flexible smart sensor systems is believed to be an added value in several domains, where smart bandages monitor the vital parameters of senior citizens and professional sportsmen and where smart flexible RFID tags monitor the quality of our food supply. The existing organic electronics technologies on foil provide a broad range of opportunities to build sensors of all kinds, e.g. temperature sensors, chemical sensor, pressure sensors and pH sensors. They all make use of the chemical or physical properties of the pentacene material or of the other layers in the available technologies. What is more, the variability of behavioral parameters, which is predominantly experienced as harmful, is often turned into the core functionality in these sensors. Nevertheless it is true that this domain is more difficult than it seems to be and that all those future applications are still far from an implementation into real products.

In this chapter first a glance was taken at the application field and attention was spent to the requirements of sensors in these applications. Most parameters in the human vital functions and in the food industry change rather slowly. Therefore, depending on the specific application, an input bandwidth of only a few Hertz to a few tens of Hertz is sufficient for a sensor in these applications.

Subsequently the state-of-the-art was presented. Several types of sensors were reported in literature. Most of those employed the sensitivity of pentacene to environmental parameters. Besides, also a temperature sensor with transistors biased in the subthreshold region was published.

This was followed by an overview on the existing and often used touch pad architectures where both resistive and capacitive touch sensors were elucidated.

Then the design of a flexible $1D$ linear sensor and a flexible $2D$ touch pad was discussed. The $1D$ capacitive sensor had a simulated linear accuracy of 2 mm on a 4 cm line. The 4×4 $2D$ sensor reached a simulated accuracy of 0.3 mm after interpolation on a 3.5×3.5 cm^2 area. The 3 dB frequency of the sensor was 1.5 kS/s hence a frame rate of 93 Hz was reached.

Finally the obtained measurement results were further discussed and further improvements for a redesign were formulated.

References

Kawaguchi H, Someya T, Sekitani T, Sakurai T (2005) Cut-and-paste customization of organic FET integrated circuit and its application to electronic artificial skin. IEEE J Solid-State Circuits 40(1):177–185

He DD, Nausieda IA, Ryu KK, Akinwande AI, Bulovic V, Sodini CG (2010) An integrated organic circuit array for flexible large-area temperature sensing. IEEE International in Solid-State Circuits Conference Digest of Technical Papers (ISSCC), pp 142–143

Zan HW, Tsai WW, Lo YR, Wu YM, Yang YS (2011) Pentacene-based organic thin film transistors for ammonia sensing. IEEE Sensors J 99:1

Mori T, Kikuzawa Y, Noda K (2009) Improving the sensitivity and selectivity of alcohol sensors based on organic thin-film transistors by using chemically-modified dielectric interfaces. IEEE in Sensors pp 1951–1954

Bo L, Xie G-Z, Du X-S, Xian L, Ping S (2009) Pentacene based organic thin-film transistor as gas sensor. International conference in apperceiving computing and intelligence analysis (ICACIA), pp 1–4

Subramanian V, Lee JB, Liu VH, S. Molesa (2006) Printed electronic nose vapor sensors for consumer product monitoring. IEEE international in solid-state circuits conference (ISSCC) digest of technical papers, pp 1052–1059

Darlinski Grzegorz, Bottger Ulrich, Waser Rainer, Klauk Hagen, Halik Marcus, Zschieschang Ute, Schmid Gunter, Dehm Christine (2005) Mechanical force sensors using organic thin-film transistors. J Appl Phys 97(9):093708

Manunza I, Sulis A, Bonfiglio A (2006) Pressure sensing by flexible, organic, field effect transistors. Appl Phys Lett 89: 143502-1–143502-3

Woo JS, Wook JJ, Seongpil C, Youl KS, Ik CK, BK Ju (2010) The vertically stacked organic sensor-transistor on a flexible substrate. Appl Phys Lett 97(25):253309

Lo HW, Tai Y-C (2007) Characterization of parylene as a water barrier via buried-in pentacene moisture sensors for soaking tests. 2nd IEEE international conference on nano/micro engineered and molecular systems (NEMS '07), pp 872–875

Caboni A, Orgiu E, Barbaro M, Bonfiglio A (2009) Flexible organic Thin-Film transistors for pH monitoring. Sensors J IEEE 9(12):1963–1970

Kim KH, Lee S-G, Han J-E, Kim T-R, Hwang S-U, Deok AS, You I-K, Cho K-I, Song T-K, Yun K-S (2009) Transparent and flexible tactile sensor for multi touch screen application with force sensing. International conference in solid-state sensors, actuators and microsystems, Transducers, pp 1146–1149

Bai Y-W, Chen C-Y (2007) Using serial resistors to reduce the power consumption of resistive touch panels. IEEE international symposium on in consumer electronics (ISCE). pp 1–6

Kim H-K, Lee S-G, Han J-E, Kim T-R, Hwang S-U, Ahn SD, You I-K, Cho K-I, Song T-K, Yun K-S (2009) Transparent and flexible tactile sensor for multi touch screen application with force sensing. International conference in solid-state sensors, actuators and microsystems, transducers. pp 1146–1149

Wang T-M, Ker M-D (2011) Design and implementation of capacitive sensor readout circuit on glass substrate for touch panel applications. International symposium on VLSI design, automation and test (VLSI-DAT). pp 1–4

Wilson TV (2007) How the iPhone works. http://electronics.howstuffworks.com/iphone.htm,

Yokota T, Sekitani T, Tokuhara T, Zschieschang U, Klauk H, Huang T-C, Takamiya M, Sakurai T, Someya T (2011) Sheet-type organic active matrix amplifier system using Vth-Tunable Pseudo-CMOS circuits with floating-gate structure. Dec. 2011

Gelinck Gerwin H, Edzer HA, Huitema, Erik van V, Cantatore E, Schrijnemakers L, Jan BPH van der Putten, Tom Geuns CT, Beenhakkers M, Jacobus BG, Huisman B-H, Eduard JM, Benito EM, Fred JT, Albert WM, Bas JE van Rens, Dago de Leeuw M (2004) Flexible active-matrix displays and shift registers based on solution-processed organic transistors. Nat Mater 3(2):106–110

Gelinck GH, van Veenendaal E, Coehoorn R (2005) Dual-gate organic thin-film transistors. Appl Phys Lett 87(7):073508

Marien H, Steyaert M, Veenendaal EV, Heremans P (2012) 1D and 2D analog 1.5 kHz air-stable organic capacitive touch sensors on plastic foil. IEEE international conference in solid-state circuits digest of technical papers (ISSCC)

Lapinski M, Domaradzki J, Prociow EL, Sieradzka K, Gornicka B (2009) Electrical and optical characterization of ITO thin films. International workshop in students and young scientists photonics and microsystems. pp 52–55

Halik M, Klauk H, Zschieschang U, Kriem T, Schmid G, Radlik W, Wussow K (2002) Fully patterned all-organic thin film transistors

University of Tokyo (2012) http://www.ntech.t.u-tokyo.ac.jp, 08/03/2012

Chapter 6
DC-DC Conversion

The interest for integrated DC-DC converters is greater than ever due to the increasing gap between the battery voltage and supply voltage levels required for state-of-the-art chip design. Moreover the multiple-voltage system design strategy, often applied for power optimization in large systems, also requires the generation of separate voltage levels. In organic electronics technologies the DC-DC converters are new. In these immature technologies several techniques are needed in order to increase the quality, i.e. gain, reliability, and other specifications, of all kinds of circuits. The direct application field of DC-DC converters is not yet fully known. Nevertheless in this dissertation the demand for DC-DC conversion becomes evident in Sects. 3.3.3 and 4.3.1, where DC bias voltages, respectively above the supply voltage and below the ground voltage, were applied for biasing gates and backgates of transistors. These bias voltages improve the circuit reliability. The goal of the integrated DC-DC converter is of course the generation of these required bias voltages on the chip and in a power-efficient way. This chapter covers the implementation and the integration of a DC-DC converter in organic electronics technology and the resulting improvement of the reliability of other circuits.

In Fig. 1.5 the block diagram of an organic smart sensor system was presented. In this chapter the DC-DC converter building block is elucidated. In Sect. 6.1 the application field for organic DC-DC converters is investigated and design specifications are derived. Section 6.2 systematically discusses the applicability of several DC-DC converter architectures. Then three DC-DC converter designs and their corresponding measurement results are presented in Sect. 6.3. Section 6.4 provides the discussion of the results, especially of the efficiency and the applicability of the presented designs. Finally this chapter is concluded in Sect. 6.5.

H. Marien et al., *Analog Organic Electronics*, Analog Circuits and Signal Processing,   129
DOI: 10.1007/978-1-4614-3421-4_6, © Springer Science+Business Media New York 2013

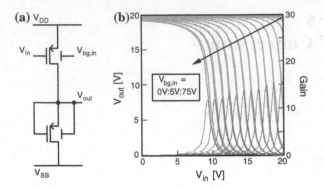

Fig. 6.1 **a** An inverter with a zero-V_{GS} pull-down transistor and a pull-up transistor with a backgate bias pin $V_{bg,in}$. **b** Bias voltages up to 75 V applied to $V_{bg,in}$ increase the noise margin of an inverter significantly (Myny et al. (2011))

6.1 Application Field

The application field for organic DC-DC converters is twofold.

In the first place there is the availability of a technique for tuning a transistor's V_T in the applied technology through the backgate that significantly increases the quality of both digital and analog circuits. This technique requires voltages above V_{DD}. A detailed discussion about the technique is given in 6.1.1.

In the second place a bias voltage below V_{SS} is required for biasing transistors in the depletion region and making them behave like resistors. This technique to bias transistors in the depletion region is discussed in Sect. 6.1.2.

6.1.1 Dual-V_T Technology

The applied organic electronics technology provides the use of backgates. They are physically located on top of the transistors and influence the V_T of the underlying transistor. This so-called backgate steering technique was elaborated on in Sect. 2.4.2.1 A very important advantage of backgate steering is that this influence of a backgate voltage on the V_T of the transistor is very linear. In Myny et al. (2011) the backgate steering technique was used for tuning the V_T of the pull-up transistors in digital gates. As can be seen in Fig. 6.1b bias voltages above the supply voltage strongly improve the noise margin of the inverter with an optimum around 20 V for a V_{DD} of 20 V. This optimum depends on the relative W/L sizing between pull-up and pull-down transistors, which is 1 : 1 in this case.

This same technique, called threshold voltage tuning (TVT), enables some freedom in the choice of the DC levels of an analog differential amplifier which facilitates the DC connection of consecutive amplifiers. A drawback of the single-stage differential amplifier discussed in Sect. 3.3.1 is that the DC levels of input and output nodes are different. Therefore an additional block is required when connecting two single-stage amplifiers together. The most efficient way to connect successive amplifiers is through high-pass filters. These high-pass filters can be left out by applying the back-gate steering technique to the transistors of the differential input pair, as explained in Sect. 3.3.3. Again this technique demands for a bias voltage above V_{DD}. Here a 30 V. to 40 V. bias voltage is required for a 15 V. supply voltage.

6.1.2 Depleted Transistors

So far, organic electronics technology has been optimized for digital applications. The availability of p-type transistors and capacitors as building blocks is sufficient for the design of organic flexible displays. In this work analog circuits are investigated which often require resistive components. However, at the moment no specific layers for making resistors are available in the organic electronics technology yet. Nevertheless transistors can be biased in the depletion region where they acquire a linear V_{SD} − I_{SD} characteristic. The conditions for this linear behavior, namely V_{SG} being large and much larger than V_{SD}, is fulfilled by applying a low bias voltage to the gate. This bias voltage must be much lower than V_{SS} in order to guarantee the resistor to be rail-to-rail applicable. The circuits presented in Sects. 3.3.2 and 4.3.1 apply resistive components. The lowest bias voltage, such a resistive transistor can bear, is around −40 V. For lower bias voltages the transistor would break down or bias stress would degrade its behavior. Therefore a negative bias voltage generated in a DC-DC converter should reach down to −40 V.

6.1.3 Specifications

The overview of the application field enables the extraction of specifications that organic DC-DC converters should meet in order to improve organic circuit behavior. The digital dual-V_T application in 6.1.1 mentions output voltages up to 75 V. It must be said that this value is not a hard specification as it is connected to a 1 : 1 ratio between pull-up and pull-down transistors. For other ratios a more moderate voltage is sufficient. In the analog applications a bias voltage in the range of 30 V to 40 V is required. This voltage is used as the specification for the high output voltage of the converter designs presented in this chapter. Similarly as the specification for the low output voltage a value of −40 V is recommended.

In all applications the converter is used for biasing gates and backgates that do not draw direct current except for a marginal leakage current. This leakage current is estimated for a minimal transistor to be less than 0.1 pF. Since the most complex organic circuits known at the moment consist of $1,000 - 10,000$ transistors, an output current of 1 n A is sufficient to drive such a circuit. This 1 n A output current is used

Table 6.1 Overview of the minimal design specifications for the organic DC-DC converters in this work

Specification	Value	
	High voltage output	Low voltage output
V_{out}	30–40 V	−40 V
I_{out}	1 nA	1 nA

as a specification for the converters in this chapter. The design specifications for an organic DC-DC converter are summarized in Table 6.1.

6.2 Topology

Nowadays a lot of research is focusing on integrated DC-DC converters in standard Si CMOS technology (Breussegem and Steyaert 2010; Wens et al. 2007). The point of interest is to let these integrated converters efficiently transform electrical power from the supply voltage level to other DC voltage levels by switching reactive components in different positions. The DC-DC converters can be subdivided in two functional families, i.e. up-converters and down-converters. Down-converters generate voltage levels that are situated between the ground and the supply voltage. Up-converters on the other hand are converters that generate voltage levels that are located outside the ground-to-supply voltage range. Both are discussed in Sect. 6.2.1. Furthermore DC-DC converters can also be subdivided in two main architectural families, i.e. inductive converters and capacitive converters. Inductive converters apply inductors together with switches in order to generate the required voltage level whereas capacitive converters apply capacitors. Inductive converters, as well as the reason why they are not applicable in organic electronics technology nowadays, are dealt with in Sect. 6.2.2. The family of capacitive converters provides several architectures that at first glance are applicable for organic electronics technology. These architectures all have their particular switching scheme. The applicability in organic electronics technology of the series-parallel, voltage doubler, Fibonacci and Dickson architectures is investigated in Sect. 6.2.3.

6.2.1 Up- and Down-Conversion

DC-DC converters can intuitively be divided in up-converters that generate a voltage higher than the supply voltage, and down-converters that convert power from the supply voltage to a voltage between the supply voltage and the ground level. A more mathematical approach to both types of converters and their specifications is presented in the following sections.

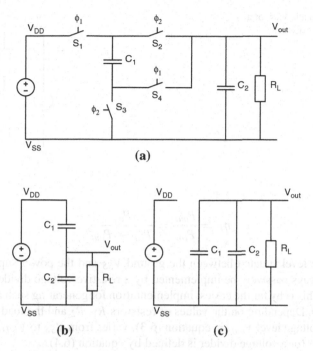

Fig. 6.2 **a** A capacitive series-parallel down-converter. **b** The converter in phase ϕ_1. **c** The converter in phase ϕ_2

6.2.1.1 Down-Converters

DC-DC down-converters apply reactive components, either capacitors or inductors, in a schematic with switches for creating their outputs. By connecting the reactive components in different positions an output power P_{out} is produced at an output voltage V_{out}. Figure 6.2 gives the example of a capacitive series-parallel down-converter that generates an output voltage of $V_{DD}/2$. In clock phase ϕ_1 the capacitors are connected in series through switches S_1 and S_4 and they are charged to $V_{DD}/2$ each by the supply voltage V_{DD}. In clock phase ϕ_2 the capacitors are connected in parallel through switches S_2 and S_3. They are both still charged to $V_{DD}/2$ and provide this voltage to the output of the converter.

The conversion ratio k_{out} is defined in equation (6.1). For down-converters this ratio is always between 0 and 1. The power efficiency η_P for a certain output voltage is defined by equation (6.2) where P_{tot} is the total power dissipated in the circuit and P_{int} the internally dissipated power.

$$k_{out} = \frac{V_{out}}{V_{DD}} \tag{6.1}$$

Fig. 6.3 Schematic view of a resistive voltage divider

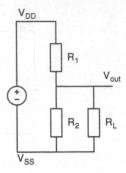

$$\eta_P = \frac{P_{out}}{P_{tot}} = \frac{P_{out}}{P_{int} + P_{out}} \tag{6.2}$$

A voltage level situated between the ground V_{SS} and the power supply voltage V_{DD} can always passively be implemented by a resistive voltage divider, as shown in Fig. 6.3. This is by far the easiest implementation for generating such a DC output voltage level. Depending on the values of resistors R_1, R_2 and the load resistor R_L the output voltage level V_{out}, in equation (6.3), varies from V_{SS} to V_{DD}. The power efficiency η_P for a voltage divider is defined by equation (6.4).

$$V_{out} = \frac{R_2 \parallel R_L}{R_1 + R_2 \parallel R_L} \approx \frac{R_2}{R_1 + R_2} \quad (with \ R_L \gg R_2) \tag{6.3}$$

$$\eta_P = \frac{P_L}{P_{tot}} = \frac{P_L}{P_1 + P_2 + P_L} \quad (with \ P_i = V_i \cdot I_i \ for \ i = 1, 2, L) \tag{6.4}$$

where P_i is the power dissipated in each resistor R_i, V_i and I_i the voltage over and the current through each resistor R_i. The most beneficial situation is obtained when all the current flows through R_L and no current at all flows through R_2, i.e. when R_2 is infinite. Then $\eta_{P,res}$ is defined by equation (6.5). From this result it can be concluded that a voltage divider has a poor η_P, especially for a low k_{out}.

$$\eta_{P,res} = \frac{V_{out}}{V_{DD}} = k_{out} \tag{6.5}$$

Another drawback of the voltage divider is the value of its output resistance of the equivalent circuit $R_{out,eq}$. Resistors R_1 and R_2 have already been determined in order to optimize η_P and no measure of freedom remains. Especially for V_{out} lower than $V_{DD}/2$, $R_{out,eq}$ equals R_L and becomes even larger. Then any change in the value of R_L, i.e. in the case of a variable load, has a tremendous impact on V_{out}. Both the poor η_P and the high $R_{out,eq}$ demand for more efficient power conversion techniques that can be found in DC-DC down conversion.

The efficiency of a resistive voltage divider forms a lower boundary for the applicability of a down-converter. As long as the active down-converter's efficiency does not climb above the efficiency of the voltage divider, discussed in the highlighted box on page 134, it should better be replaced by that voltage divider. A down-converter core only consists of reactive components and switches. Assuming the switches and the reactive components to be ideal, these converters do not dissipate internal power. This means that their η_P, defined in equation (6.2), reaches 1 independently of the conversion ratio k_{out}. Therefore down-converters are of interest especially for low k_{out}. A good measure of the quality of a down-converter is the difference between η_P and k_{out}, which corresponds to the optimal efficiency $\eta_{P,res}$ of the voltage divider. This measure is called the efficiency enhancement factor (EEF) (Wens & Steyaert 2011).

6.2.1.2 Up-Converters

DC-DC up-converters, just like down-converters, apply reactive components, either capacitors or inductors, in a schematic with switches for creating their outputs. The reactive components are now switched in such a way that an output power P_{out} is produced at an output voltage V_{out} that is situated above V_{DD} or below V_{SS}. This second case ($V_{out} < V_{SS}$) is fully dual to $V_{out} > V_{DD}$ and is no more separately mentioned in this section. Up-converters are characterized by their conversion ratio k_{out} always greater than 1 and by their power efficiency η_P. Unlike for down-converters, there is no resistive alternative to generate V_{out} and consequently the measure of quality of a down-converter can not be extrapolated towards up-converters.

6.2.2 Inductive Converters

The core element of an inductive DC-DC converter is an inductor. In this section the principles of inductive converters are briefly discussed. Figure 6.4a shows the schematic view of an ideal inductive DC-DC boost converter, which is an up-converter. The operation mode consists of two clock phases that are visualized in Figs. 6.4b and 6.4c. During the charging phase ϕ_1 the current through the inductor increases linearly and energy is stored in the magnetic field of the inductor. In the discharging phase ϕ_2 the energy stored in the inductor during ϕ_1 is released and current flows to the capacitor that is further charged. The mathematical approach to the behavior of this boost converter goes beyond the scope of this work.

Lately inductive converters have been integrated in Si CMOS technology (Wens et al. 2007). This full integration has been delayed by the dimensions and the cost of chip area in Si CMOS and this is also the case in organic electronics technology. The required dimensions obstruct the use of inductors for DC-DC conversion in this

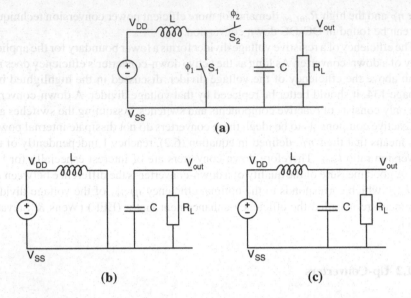

Fig. 6.4 **a** Schematic view of an ideal boost converter. **b** The converter in phase ϕ_1. **c** The converter in phase ϕ_2

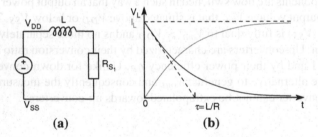

Fig. 6.5 Charging behavior of a series RL chain

work. The inductance that is required for implementation in the boost converter in an organic electronics technology with a 10 kHz bandwidth can be roughly estimated. During the charging phase ϕ_1 the inductor is charged through switch S_1 as can be seen in Fig. 6.4. In the applied technology this switch is not ideal and a series resistor is present. As a start this resistance is optimistically estimated around 10 kΩ. The charging curve of the inductor is shown in Fig. 6.5. For optimal efficiency the charging of the inductor should be in the linear part of the curve which means that the time constant τ_{LR} has to equal the clock period which is 0.1 ms, as written in equation (6.6):

$$\tau_{LR} = \frac{L}{R_{S1}} = \frac{1}{f_p} \tag{6.6}$$

where R_{S1} is the switch resistance and f_p the bandwidth of the technology. From this equation the required inductance of $1\,H$ is calculated. This gigantic inductance value is reached by a 700 turns, 1 m diameter inductor. It is clear now that several orders of magnitude have to be scaled down before inductive DC-DC converters can be considered in small integrated circuits.

6.2.3 Capacitive Converters

The key component in capacitive DC-DC converters is a capacitor. This capacitor temporarily stores an amount of electrical energy that is then converted to a higher or lower voltage. Capacitive converters are not limited by the resistance of their switches. In every clock cycle an amount of energy flows from one to another capacitor. The clock cycle should be long enough to fully charge every capacitor. A very high switching resistance now only implies a lower clock frequency, hence less energy packets are delivered to the output or, better said, the output power of the converter is lower. Since the application field only demands for high voltages for biasing gates and backgates rather than for power delivery, capacitive converters can fulfill these needs. In this section only the capacitive up-converters are discussed since down converters are not directly of interest in this work. Four different architectures are investigated for their implementation in organic electronics technology.

6.2.3.1 Series-Parallel Architecture

The series-parallel up-converter architecture, presented in Fig. 6.6, is very similar to the series-parallel down-converter architecture from Fig. 6.2. In the charging phase ϕ_1 capacitor C_1 is charged by the supply voltage V_{DD} through switches S_1 and S_2 while capacitor C_2 provides power to R_L. In the discharging phase ϕ_2 capacitor C_1 is now connected to V_{DD} through switch S_3 creating a voltage level of $2.V_{DD}$ over C_2 through S_4. The working principle of this converter can be extrapolated towards architectures with more stages. The ideal output voltage of an nth order series-parallel converter is $(n + 1).V_{DD}$.

In order to meet the specifications for a DC-DC converter, discussed in Sect. 6.1, a 3-stage converter is required that is presented in Fig. 6.7. The ideal output voltage of this 3-stage series-parallel converter is $4.V_{DD}$. The most difficult part of integrating these converters in organic electronics technology, is the implementation of their active elements: the switches. These can only be implemented as the p-type transistors that are available in the applied technology. In this paragraph all the switches of the 4-stage series-parallel converter are examined.

Switches $S_{1,i}$ conduct current during ϕ_1 and are always connected with one pin to V_{DD}. The voltage at their other pin, node n_i, switches between V_{DD}, during ϕ_1, and $(i + 1).V_{DD}$, during ϕ_2. $S_{1,i}$ is closed during ϕ_1, hence the nodes n_i are at V_{DD}. While the nodes n_i are high no current may flow through $S_{1,i}$. These switches

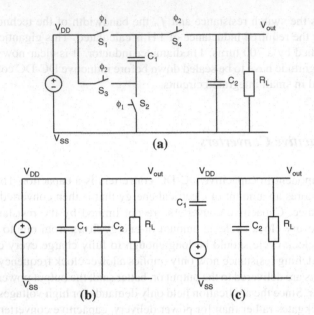

(a)

(b)   (c)

Fig. 6.6 a Schematic of an ideal series-parallel up-converter. **b** The schematic during ϕ_1. **c** The schematic during ϕ_2

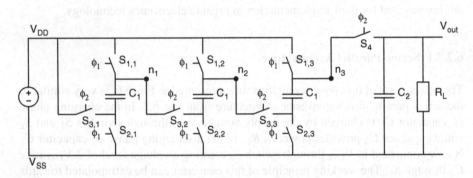

Fig. 6.7 Schematic of an ideal 3-stage series-parallel up-converter

can be passively implemented as diode-connected transistors and do not require an additional clock signal which is beneficial for the design. Switches $S_{2,i}$ also conduct current during ϕ_1 and are always connected to V_{SS} with one pin. Their other pin is also connected to V_{SS} during ϕ_1 but during ϕ_2 to n_{i-1}. During ϕ_2 these nodes n_{i-1} are at a high voltage $i.V_{DD}$. No passive implementation can be applied to these switches. Contrarily an additional clock signal is required for each of the 3 switches $S_{2,i}$. For switch $S_{2,1}$ this can be a standard switching signal between V_{DD} and V_{SS} whereas switches $S_{2,2}$ and $S_{2,3}$ require respective voltage levels of $2.V_{DD}$ and $3.V_{DD}$ to be switched off. This severely complicates the implementation of this topology.

Table 6.2 Overview of the implementation details of switches in the 3-stage series-parallel topology. The required number of each type of switch is given. The total number of switches and the total number of clock signals needed are also included

	Passive	DC bias voltage	Clocked $\leqslant V_{DD}$	Clocked $> V_{DD}$	#S_i	#Clock signals
3-st. Series-Parallel	4	2	2	2	10	4

Switches $S_{3,i}$ are closed during ϕ_2. During ϕ_1 they are all biased with V_{DD} at their left plate and with V_{SS} at their right plate. During that phase $S_{3,i}$ may not draw current hence a bias voltage of V_{DD} or higher is required. During ϕ_2 both plates of $S_{3,i}$ are at the same voltage $i.V_{DD}$. Except for $S_{3,1}$, all $S_{3,i}$ can be switched by a DC voltage V_{DD} or slightly higher. These switches are of no cost for the implementation of this converter. Switch $S_{3,1}$ cannot be biased with a DC voltage. Nevertheless, it can be biased pretty easily with a standard clock signal between V_{DD} and V_{SS} and does not present a problem for the implementation.

Finally switch S_4 is in a very similar situation to switches $S_{1,i}$ and can be implemented in the same way, i.e. passively with a diode-connected transistor.

The conclusion is that most of the switches, i.e. $S_{1,i}$, $S_{3,i}$ and S_4, can be easily implemented. However, switches $S_{2,i}$ make the implementation of this topology very difficult since they require high swing switching signals at voltage levels far beyond V_{DD}. Table 6.2 summarizes the numbers of switches of different types and the number of clock signals needed for this topology.

6.2.3.2 Voltage Doubler Architecture

The schematic of a single-stage voltage doubler is similar to that of the series-parallel architecture in Fig. 6.6 and, as a result, the same working principle applies. The difference between the voltage doubler and the series-parallel architecture lies in the connection between consecutive stages. Figure 6.8a shows a 2-stage voltage doubler architecture. In the circuit a capacitor C_{o1} is present that stores the $2.V_{DD}$ output of the first stage at node n_{o1}. This DC voltage is then used as the input voltage for the subsequent voltage doubler. The output voltage V_{out} at n_{out} has an ideal value of $4.V_{DD}$. The charging principles of this converter during ϕ_1 and ϕ_2 are visualized in Figs. 6.8b and 6.8c. More generally the output of an nth order voltage doubler is 2^n. This topology reaches the same output voltage as the series-parallel topology with only two instead of three stages. Please note that this topology applies two more capacitors that must be large to reduce the output ripple.

The switches in a voltage doubler behave very comparably with those in the series-parallel converter, discussed in 6.2.3.1. Switch $S_{1,1}$ must conduct forward current while both its pins are at a voltage V_{DD} and must not draw backward current when node n_1 is at $2.V_{DD}$. $S_{4,1}$ must block backwards current when n_1 is at V_{DD} and

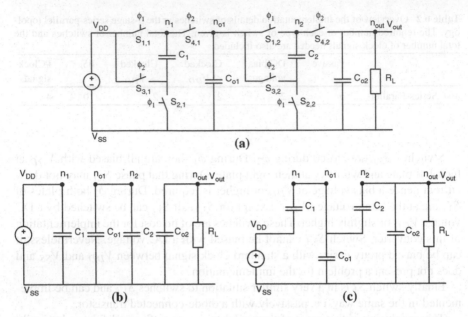

Fig. 6.8 **a** Schematic of an ideal 2-stage voltage doubler. **b** The schematic during ϕ_1. **c** The schematic during ϕ_2

Table 6.3 Overview of the implementation details of switches in the 2-stage voltage doubler topology. The required number of each type of switch is given. The total number of switches and the total number of clock signals needed are also included

	Passive	DC bias voltage	Clocked $\leqslant V_{DD}$	Clocked $> V_{DD}$	#S_i	#Clock signals
2-st. Voltage doubler	4	0	4	0	8	4

conduct when both its pins are at $2.V_{DD}$. A passive implementation is possible for both $S_{1,1}$ and $S_{4,1}$.

Switch $S_{2,1}$ conducts current when its pins are brought to the ground level V_{SS} whereas it is open when the upper pin is at V_{DD}. Here a clock signal between V_{DD} and V_{SS} is required. The inverse clock signal is needed for $S_{3,1}$ that is connected to V_{DD} with one pin and conducts oppositely to $S_{2,1}$. Since the capacitor C_{o1} stores the intermediate $2.V_{DD}$ voltage at node n_{o1}, this voltage can be seen as the input voltage for the second voltage doubler where it can also serve as a rail for generation of clock signals. Therefore the discussion for switches $S_{1-4,2}$ is identical to the discussion of $S_{1-4,1}$ where V_{DD} is everywhere replaced by $2.V_{DD}$.

The conclusion for this topology is more optimistic than for the series-parallel architecture since the required additional clock signals can be easily implemented. Table 6.3 summarizes the analysis of the voltage doubler implementation.

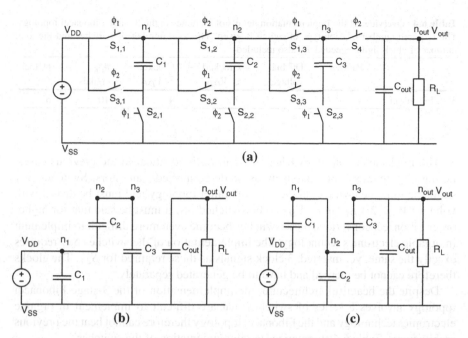

Fig. 6.9 **a** Schematic of the ideal 3-stage Fibonacci up-converter. **b** The schematic during ϕ_1. **c** The schematic during ϕ_2

6.2.3.3 Fibonacci Architecture

The 1-stage series-parallel, the voltage doubler and the Fibonacci architectures all have the same schematic, shown in Fig. 6.6. The difference in behavior between the architectures relates to the connection between consecutive stages. The beauty of this architecture is the correspondence of the output voltage with the Fibonacci series[1] $Fib(n + 3)$. In Fig. 6.9a the schematic view of a 3-stage Fibonacci up-converter architecture is shown. Note that in this topology the clock phase of the consecutive stages is inverted. This is visualized in Figs. 6.9b and 6.9c where the schematic view of the Fibonacci converter is drawn for clock phases ϕ_1 and ϕ_2. During ϕ_1 and ϕ_2 respectively the odd capacitors, C_1 and C_3, and the even capacitors, C_2 and C_{out}, are charged. Every capacitor is charged by its two preceding capacitors which results in the Fibonacci-like increasing output voltage for increasing number of stages in the circuit. The output voltage of the ideal 3-stage Fibonacci topology is $5.V_{DD}$. Generally in an ideal n-stage Fibonacci converter v_{out} equals $Fib(n + 3)$.

---

[1] L. Fibonacci (°1,170-†1,250) was an Italian mathematician who in his book *Liber Abaci* introduced the Arabic numerals in Europe. He is most famous for the Fibonacci numbers where every number $Fib(n)$ equals the sum of $Fib(n - 1)$ and $Fib(n - 2)$, where $Fib(0)$ is 0 and $Fib(1)$ is 1. These numbers were part of an ideal mathematical model for the population growth of rabbits. Fibonacci numbers pervade in several natural environments, i.e. the number of petals in a flower.

Table 6.4 Overview of the implementation details of switches in the 3-stage Fibonacci topology. The required number of each type of switch is given. The total number of switches and the total number of clock signals needed are also included

	Passive	DC bias voltage	Clocked $\leqslant V_{DD}$	Clocked $> V_{DD}$	#S_i signals	#Clock
3-st. Fibonacci	4	0	2	4	10	6

The implementation of switches $S_{1,i}$ and S_4 is identical to the previous cases, i.e. they are implemented passively as diode-connected transistors. Switches $S_{2,i}$ show the same behavior as in the series-parallel topology and must be driven with voltages V_{DD}, $2.V_{DD}$ and $3.V_{DD}$ to be switched off. It must be said that for higher order Fibonacci converters, these switches become even more difficult to implement in organic electronics technology. The implementation of the switches $S_{3,i}$ requires exactly the same, yet inverted, 3 clock signals as those required for $S_{2,i}$. The clocks therefore cannot be merged and have to be generated separately.

Despite the beautiful architecture, the implementation of the 3-stage Fibonacci topology involves a series of switches that are difficult to implement in organic electronics technology and the Fibonacci topology therefore cannot beat the previous architectures. Table 6.4 summarizes the implementation of the switches.

6.2.3.4 Dickson Architecture

The Dickson (Dickson 1976) up-converter architecture is another topology of the capacitive up-converters family. Figure 6.10a shows a 3-stage implementation of this topology. The lower plates of capacitors C_1 and C_3 are connected to node n_4 and are driven by the same set of switches, which has a positive influence on the complexity of the implementation. Moreover nodes n_4 and n_5 are never driven to a voltage level above V_{DD}, which facilitates the generation of the switching signals. The working principle of this topology is visualized in Figs. 6.10b and 6.10c. During ϕ_1 and ϕ_2, respectively, the odd and even capacitors are charged with the voltage over their preceding capacitor incremented with V_{DD}, hence every capacitor C_i is charged with $i.V_{DD}$. Correspondingly a voltage of $(n+1).V_{DD}$ is obtained at the output of an n-stage converter, which means $4.V_{DD}$ for the case of the 3-stage implementation.

In order to meet the specifications deduced in Sect. 6.1 a 3-stage Dickson converter is required. Just like in the previous architectures, here also the switches $S_{1,i}$ and S_4 can be implemented passively. Biasing the more difficult switches to bias, i.e. $S_{2,i}$ and $S_{3,i}$, now only requires two clock signals that switch between V_{DD} and V_{SS}. In this way the implementation of the switches becomes a lot easier than in the previous architectures.

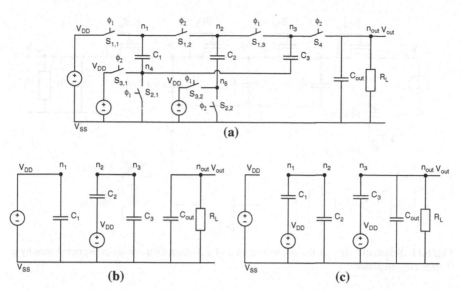

Fig. 6.10 a Schematic of the ideal 3-stage Dickson up-converter. **b** The schematic during ϕ_1. **c** The schematic during ϕ_2

Table 6.5 Comparative overview of the implementation details of switches in the series-parallel, the voltage-doubler, the Fibonacci and the Dickson architectures. The required number of each type of switch is given. The total number of switches and the total number of clock signals needed are also included

	Passive	DC bias voltage	Clocked $\leqslant V_{DD}$	Clocked $> V_{DD}$	#S_i	#Clock signals
3-st. Series-Parallel	4	2	2	2	10	4
2-st. Voltage doubler	4	0	4	0	8	4
3-st. Fibonacci	4	0	2	4	10	6
3-st. Dickson	4	0	2	0	8	2

6.2.3.5 Discussion

Table 6.5 summarizes the implementation of the series-parallel, the voltage-doubler, the Fibonacci and the Dickson architectures. The Fibonacci converter needs six additional clock signals, four of which must be above the supply voltage. Therefore this topology is not of interest. The series-parallel converter scores slightly better with only four additional clock signals, two of which must be above the supply voltage. The voltage doubler and the Dickson architectures only require clock signals that can easily be generated with the available voltage levels and perform almost equally. However, the Dickson converter only requires two of these signals, whereas the voltage doubler requires four of these. An interesting feature of the Dickson topology is that it only uses eight switches in a 3-stage converter whereas the other 3-stage

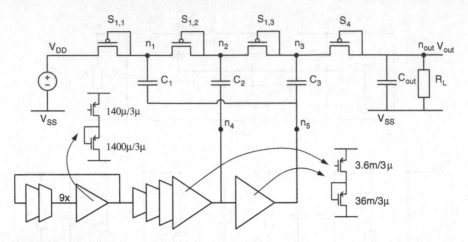

Fig. 6.11 Schematic view of the implementation of a 3-stage Dickson up-converter driven by a 9-stage ring oscillator

converters needs ten switches. Contrarily to the other architectures, in the Dickson architecture the forward path is fully isolated from switches connected to the ground, which reduces the number of parasitic paths that induce losses. In the topological discussion the Dickson converter is revealed as the best applicable topology in organic electronics technology. It is further used for the circuit design in Sect. 6.3.

6.3 Designs

In Sect. 6.2 the 3-stage Dickson up-converter architecture was selected as the best suitable topology for implementation in organic electronics technology. In this section three consecutive designs are presented together with their simulation and measurement results.

6.3.1 Design 1

6.3.1.1 Schematic

In Fig. 6.11 the first implementation of the up-converter is presented. Every Dickson stage consists of a 50 pF capacitor C_i and, as discussed in Sect. 6.2.3.4, a switch $S_{1,i}$ that is passively implemented as a diode-connected transistor with a 600 μ/5 μm W/L. S_4 is implemented identically to $S_{1,i}$ and the output capacitor C_{out} amounts to 70 pF. The pairs of switches $S_{2,1} - S_{3,1}$ and $S_{2,2}$-$S_{3,2}$ drive n_4 and n_5 with two inverted clock signals and can be easily implemented as an inverter driven by a single clock signal.

Although the passive implementation of $S_{1,i}$ and S_4 is the easiest implementation for these switches, they behave far from ideally.

Firstly, a transistor in the applied organic electronics technology is not yet fully switched off when V_{SG} is 0 V since the threshold voltage is close to 0 V. Therefore the off-current I_{off} is not very low and the I_{on}/I_{off} ratio is significantly reduced. Furthermore this ratio is very sensitive to V_T variation and V_T shift, since the current in sub-threshold changes logarithmically with V_{SG}.

Secondly, the V_{SD} of the transistor during the off state is large, around $2 \cdot V_{DD}$, which also has a small but negative influence on the I_{off} and therefore on the I_{on}/I_{off} ratio.

Thirdly, when the transistor is in the on-state after settling, all of its nodes are at the same voltage level, hence the transistor does not draw any current. During the evolution towards this final state the transistor biasing becomes worse and the R_{on} becomes higher slowing down the settling. This introduces a trade-off between clock speed and output voltage.

The clock signal is generated on the chip by the 9-stage ring oscillator implemented with inverters that each have a 140 µm/3 µm pull-up transistor and a 1,400 µm/3 µm zero-V_{GS} load transistor. In order to drive the capacitors at nodes n_4 and n_5 the clock signal is buffered with a chain of five sized inverters. The finger width of all the transistors in the ring oscillator and the buffer is reduced from the safer 5 µm towards 3 µm in order to reduce the gate-source (C_{gs}) and gate-drain (C_{gd}) capacitors with 40 %. Therefore the clock speed increases with an estimated 67 %.

6.3.1.2 Simulation

The behavior of the converter is simulated based on measurements of single transistors that are processed in the same technology. The simulated behavior of the Dickson converter with a varying power supply voltage is shown in Fig. 6.12a. The output voltage of 48 V is reached with a 15 V power supply and goes up to 88 V for a 25 V supply. The simulated clock frequency is 500 kHz, which is likely an overestimation since the simulation is done with extracted information from earlier transistor measurements and with a capacitive transistor model that poorly fits to the technology. The second curve is the simulated behavior when a load current of 10 µA is applied. Now the output voltage is lowered by 6 V and the output resistance of the converter is estimated as 0.6 MΩ.

Figure 6.12b shows the simulated behavior of nodes n_{1-3} together with the output voltage in steady-state behavior. The charging principle of this converter, discussed in Sect. 6.2.3.4, is now visualized. Node n_1 is charged by V_{DD} during ϕ_1 and then charges n_2 during ϕ_2. Node n_2 for his part charges n_3 during ϕ_1 and consequently the output node n_{out} is charged by n_3 during ϕ_2. The voltage loss, while one node

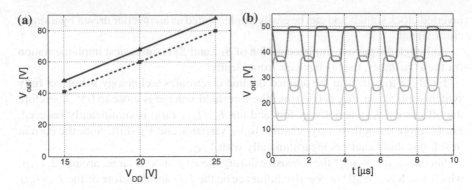

Fig. 6.12 **a** Simulated output voltage of a 3-stage Dickson up-converter without (*solid line*) and with (*dashed line*) a 10 μA load current applied. **b** Simulated steady-state transient behavior of a 3-stage Dickson up-converter output and the internal nodes n_{1-3}

is charging the other, is clearly visible and amounts to 2 V for every diode. Hence compared to the ideal Dickson behavior a voltage of 8 V is already lost at the output.

6.3.1.3 Measurement Results

The Dickson DC-DC converter is implemented on top of a *Si* wafer and all measurements are performed in a nitrogen environment in order to prevent the transistors from suffering from degradation effects. The measured clock frequency is 9 Hz and the oscillator and the buffers consume 560 μA from a 15 V power supply. Figure 6.13a shows the measured V_{out} of the 3-stage Dickson converter for a varying supply voltage V_{DD}. When only a capacitive load is applied the V_{out} reaches from 42 V for a 15 V V_{DD} up to 75 V for a 25 V V_{DD}. For higher voltages the output capacitor breaks down. The measured conversion ratio of the converter is around 3 and, moreover, since the input voltage can be arbitrarily chosen between V_{DD} and V_{SS}, a tuning range of 30 % is available. When the converter is charged with a load current of 10 μA the output voltage decreases by 15 V. Compared to the simulation results, two major differences are observed.

Firstly, the measured output V_{out} without load current applied lies 15 V lower than the simulated output. This difference is put down to more losses than the simulated voltage drop in the diode-coupled transistors, which was derived from Fig. 6.12, and to clock signals that are not fully rail-to-rail.

Secondly, the voltage drop caused by the 10 μA load current is 15 V compared to a simulated voltage drop of 7 V and correspondingly the measured output resistance is 1.2 MΩ, twice the simulated output resistance. This observation is again explained by poor modeling of the diode-connected transistors. These transistors perform both in the cut-off region and in the saturation region. Where the accuracy of the comparison

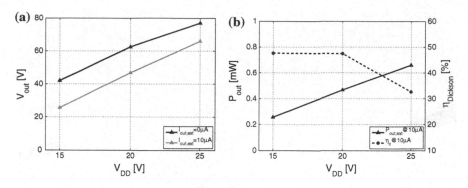

Fig. 6.13 **a** Measured DC output voltage of the 3-stage Dickson converter as a function of supply voltage V_{DD}. The black line is the output voltage for a purely capacitive load, the gray line the output voltage with a $10\,\mu A$ load current. For this measurement V_{in} is connected to V_{DD}. **b** Measured output power (*solid line*) and core power efficiency (*dashed line*) of the converter for varying supply voltage. For this measurement V_{in} is connected to V_{DD}

between two transistors biased in the same region is already limited, it even decreases further when they are biased in a different region, i.e. saturation versus cut-off.

A measure that partly explains the quality of this converter is its core efficiency, i.e. the ratio between the current at the output flowing through the load, $I_{out,ext}$, and the current at the output flowing back in the converter due to the non-ideal diodes, $I_{out,int}$.

$$\eta_c = \frac{I_{out,ext}}{I_{out,ext} + I_{out,int}} \tag{6.7}$$

$$\eta_c = \frac{V_{out}\,|_{I_{out}=10\mu A} - V_{out}\,|_{I_{out}=0\mu A}}{V_{out}\,|_{I_{out}=10\mu A} - V_{out,ideal}} \tag{6.8}$$

The core efficiency, calculated from equation (6.7), is a measure of the behavior of the converter with a specific load current. In Fig. 6.13b the output power and the core efficiency of the Dickson converter are deduced from the measured behavior in the case of a $10\,\mu A$ load current. An output power of up to $650\,10\,\mu W$ is obtained for the $25\,V$ supply voltage. The dashed curve visualizes the converter core efficiency, calculated from equation (6.8).

A maximal core efficiency of $48\,\%$ is reached in this converter. The best metric of a converter is its overall efficiency η, which is calculated from equation (6.9) and where the power of the clock generation circuit is included in the calculation. This efficiency is only around $4\,\%$ which is mainly caused by the clock driving circuit that consumes static power. This phenomenon is further discussed in Sect. 6.4.

$$\eta = \frac{P_{out,ext}\,|_{I_{out}=10\mu A}}{P_{tot}} \tag{6.9}$$

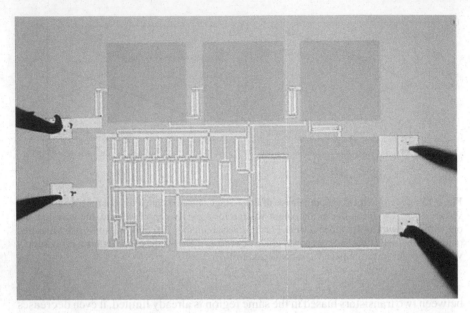

Fig. 6.14 Chip photograph of the presented Dickson DC-DC converter. This circuit is made in technology provided by Polymer Vision

Table 6.6 Summary of the measurement results for the 3-stage Dickson DC-DC converter integrated on top of a Si substrate (Marien et al. 2010)

Specification	Value
Power supply	20 V
Clock frequency	9 kHz
Voltage conversion ratio (no load)	3
Maximal output voltage (no load)	75 V
Output current	10 μ A
Output power	0.5 mW
Core efficiency	48 %
Tuning range	30 %
Oscillator and buffer current	560 μ A
Chip area	$2.8 \times 2.1\,mm^2$

The chip photograph of the Dickson DC-DC converter with a surface area $2.8 \times 2.1\,mm^2$ is shown in Fig. 6.14. On this photograph all 3 Dickson stages are visible, as well as the ring oscillators, the buffer and the output capacitor. An overview of all the measurement results is given in Table 6.6.

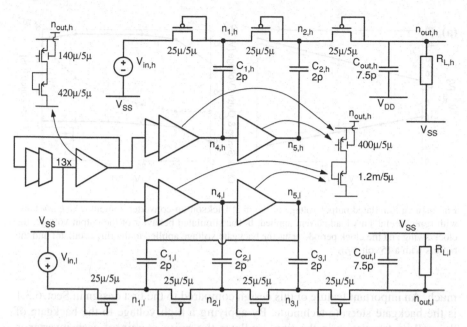

Fig. 6.15 Schematic view of the implementation of the organic dual Dickson up-converter driven by a 13-stage ring oscillator

6.3.2 Design 2

This design has been implemented in a slightly slower technology and is therefore expected to perform worse for output power than the previous design. On the other hand this technology enables the backgate steering technique that brings certain advantages at the implementation level. In this design also a low output voltage down to $-40\,\mathrm{V}$ is provided next to the high output voltage. Therefore a second Dickson converter core is included that is steered by the same clock signals. In order to reach the specifications of Sect. 6.1.3 for analog circuit improvement a 2-stage converter is applied for the high output $V_{out,h}$ and a 3-stage converter for the low output voltage $V_{out,l}$.

6.3.2.1 Schematic

Figure 6.15 shows the implementation of the second converter design. A 13-stage ring oscillator with $140\,\mu/5\,\mu$ pull-up transistors and $420\,\mu/5\,\mu$ load transistors generates the clock signals. A buffer chain drives both Dickson cores. All the diodes in the forward paths are built with $25\,\mu/5\,\mu$ transistors. Each Dickson core has a 2 pF capacitor and at the output nodes 7.5 pF capacitors are implemented. For $V_{out,h}$ this capacitor refers to V_{DD} instead of V_{SS} in order not to stress the output capacitor too

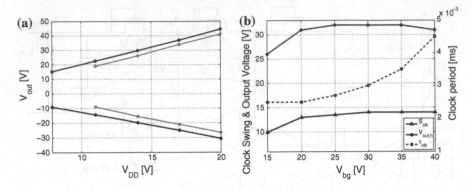

Fig. 6.16 **a** Simulated output voltage of the dual Dickson up-converter without (*black line*) and with (*grey line*) a 1 nA load current applied. **b** The simulated behavior of the output voltage, the clock swing and the clock period versus the backgate voltage applied to the ring oscillator and the buffers with a 15 volt V_{DD}

much. An important feature of this design compared to the first design in Sect. 6.3.1 is the backgate steering technique. By applying a high voltage to the backgate of the pull-up transistors in the ring oscillator the noise margin of each inverter is improved resulting in steeper clock signals with a swing that comes closer to the V_{DD} and V_{SS} rails. The latter improves the output voltage. Hence some kind of positive feedback is applied that is further investigated in Sect. 6.3.2.2. The backgate steering is also applied to the diode-connected transistors in the forward paths of the Dickson cores and improves the I_{on}/I_{off} ratio. This was demonstrated in Sect. 2.4.2.1 where both the on-state behavior and the off-state behavior of the diode were improved by employing a 4-contact diode topology. Since the specifications demand for a high output voltage but not for a high output current, the capacitors and the diodes of the converter are much smaller than in the design in Sect. 6.3.1. The capacitors have been reduced by a factor of 25 and the diodes by a factor of 40. This of course results in reduced area and power consumption.

6.3.2.2 Simulation

The simulation results of the dual Dickson up-converter are shown in Fig. 6.16a. The high output voltage that is generated through a 2-stage converter reaches 45 V for a 20 V supply voltage, whereas the low output voltage reaches down to −31 V. When a 1 nA output current is applied the high and low output voltages are reduced by 3 V and 4 V respectively. The simulated clock frequency is around 300 Hz in steady-state behavior.

The application of backgate steering in the ring oscillator and the buffers by the output voltage of the converter generates a feedback loop. During start-up this is a positive feedback loop that speeds up the start-up behavior. This is visible in

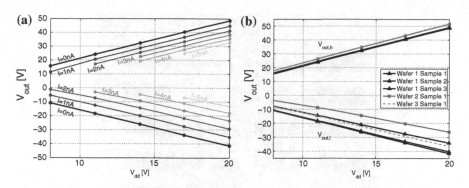

Fig. 6.17 **a** Measured high and low output voltages versus the power supply voltage V_{DD} for a varying current from 0 to 5 nA. **b** The spread in the measured behavior of the high and low output voltages for a varying V_{DD}. Each shade of gray corresponds to the measurements on a certain wafer

Fig. 6.16b where the output voltage, the clock swing and the clock period are plotted versus the backgate voltage applied. When the power supply is switched on, the ring oscillator is not perfectly biased and the clock swing is limited. The output node, which biases the ring oscillator and the buffers, is pushed to an intermediate level already. This output voltage is applied to the backgates in the ring oscillator and enhances the clock swing. This positive feedback is present until the optimum is reached at around 32 V. Then the clock period starts increasing rapidly as the pull-up transistors of the ring oscillator are pinched off. On top of that the clock swing has reached its maximum. Visibly the output voltage is no more pushed up but starts to decrease again for higher backgate voltages.

6.3.2.3 Measurement Results

This design is implemented in an organic electronics technology on foil and measurements are performed in an N_2 chamber. The measurements of both $V_{out,h}$ and $V_{out,l}$ are presented in Fig. 6.17a. The black curves correspond to the case where no current is pulled from the outputs. Then $V_{out,h}$ reaches from 16 V to 49 V for a power supply that varies between 8 V and 20 V while $V_{out,l}$ reaches from −10 to −41 V. Both outputs meet the specifications presented in Sect. 6.1.3. When load currents of 1–5 nA are applied the high and low outputs decrease by 3 V/ nA and 7 V/ nA respectively, hence both outputs have an internal resistance of 3 GΩ and 7 GΩ. The driving circuits consume 520 nA from a 20 V power supply and the measured clock speed is 300 Hz. The measured conversion ratio for $V_{out,h}$ is around 2.5 whereas this is 3 for $V_{out,l}$. Similarly to the previous design, a tuning range is available amounting to 40 % and 30 % for $V_{out,h}$ and $V_{out,l}$ respectively. The power efficiency is calculated from equation (6.10) and amounts to 0.96 % for a 20 V V_{DD} and an interpolated 2.5 nA output current. This efficiency is again very poor because of the static current

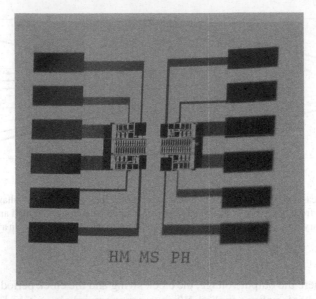

HM MS PH

Fig. 6.18 Chip photograph of the dual Dickson DC-DC converter on foil (Marien et al. 2011). On this photograph two samples of the presented circuit are visible facing eachother

consumption in the ring oscillator and the buffers. It is even slightly worse since the current in the oscillator has not decreased as much as the output current of the converter.

$$\eta = \frac{P_{out}}{P_{in} + P_{clk}} \approx \frac{P_{out}}{P_{clk}} \tag{6.10}$$

As a measure of variation in the design five samples of this circuit have been measured on three different wafers from the same batch. The result is presented in Fig. 6.17 where black, dark grey and light grey correspond to the measurements on each wafer. From the plot it can be concluded that the high output voltage $V_{out,h}$ undergoes a small variation whereas the $V_{out,l}$ suffers from a higher variation between the curves. It is expected that the variation is proportional to the number of stages in the converter. This partly explains the higher variation in the lower curves. However, on account of the symmetry between the upper and lower circuits the outlier in the lower curves is probably caused by one malfunctioning diode in the forward path as that specific output signal always measures $\sim 2/3$ of the other curves.

This converter is implemented on a foil and the active area measures $1.1 \times 1.4\,\text{mm}^2$ which is four times less than the area for the first design, although both are in different technologies hence difficult to compare. The chip photograph of this design is shown in Fig. 6.18. All measurement results are summarized in Table 6.7.

Table 6.7 Summary of the measurement results for the dual Dickson DC-DC converter integrated on a foil

Specification	$V_{out,h}$	$V_{out,l}$
Power supply	20 V	
Clock frequency	300 Hz	
Voltage conversion ratio (@20 V)	2.5	3
Maximal output voltage (@20 V)	50 V	−40 V
Output current	5 nA	−5 nA
Output power	0.17 μW	
Tuning range	40 %	30 %
Oscillator and buffer current	520 nA	
Chip area	1.1×1.4 mm^2	

6.3.3 Design 3

In order to make DC-DC converters feasible in organic electronics technology it is more important that their power consumption is minimized, rather than increasing the efficiency. The following design is a redesign of the second design where the reduction of the power consumption and the optimization of the output voltage get all the focus. Both are expected to have a positive influence on the efficiency as well.

6.3.3.1 Schematic

Figure 6.19 shows the schematic view of the converter. Three alterations as compared to the design in Sect. 6.3.2 have been introduced.

Firstly, the ring oscillator has been sized with a 3.5 times smaller W/L ratio for the transistors, which of course reduces the power consumption of the ring oscillator without drastically changing the clock frequency.

Secondly, the last buffer has been designed with another architecture. A zero-V_{GS}-load inverter, as applied in the ring oscillator, only passively pulls down the signals and moreover puts an intrinsically heavy capacitive load to the output node. Therefore such a buffer scores bad for driving a certain capacitive load. The implementation of the buffer in the improved design applies an active signal to the load transistor. This inverted input signal is easily obtained from the previous nodes in the buffer chain and the buffer can now be smaller while driving a larger capacitor at its output. The reason why this active-load inverter is not used in the ring oscillator is that in every stage the lower voltage level of the clock signal is incremented by a voltage around V_T, hence the ring oscillator would undergo a certain voltage drop per stage and in the end stop oscillating.

The third alteration that has been made is the implementation of the last diode-connected transistor S_4 in the forward path of the converter as this improves the biasing of this diode. All the diodes in the forward path are performing in a poor

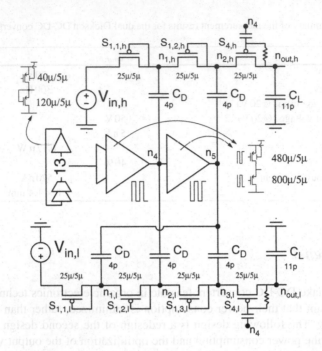

Fig. 6.19 Schematic view of the implementation of the improved organic dual Dickson up-converter driven by a 13-stage ring oscillator

biasing working region. Figure 6.20 shows the bias levels in the schematic of both the diode and the diode with high-pass filter. In the on-state the diode voltage drop decreases towards 0 V but the closer it gets to the ideal value, the closer this diode comes to the off-state, hence the behavior becomes slower. Moreover in the off-state the voltage drop over the diode, V_{DD}, is large. Therefore a certain current leaks away through the inversely biased diode that is determined by the actual value of V_T and the $V_D - I_D$ behavior of the transistor. By adding a high-pass filter to the diode S_4 the transistor biasing drastically improves. While the transistor is in the on-state the V_G values $V_{out} - V_{DD}/2$ while V_S and V_D are $V_{out} + \Delta V$ and V_{out} respectively, hence the $V_{DD}/2$ voltage drop over V_{SG} improves the on-state behavior. Similarly in the off-state the gate is now biased with $V_{out} + V_{DD}/2$ and the transistor is therefore far within the cut-off region where the current decreases exponentially for decreasing $V_{SG} < 0 V$. So the leakage current is much lower than in the passive implementation.

6.3.3.2 Simulation

To compare designs 2 and 3, a simulation of their relative current consumption has been run. The effect of reducing the transistor sizes and improving the buffer results in a current consumption that is 2.6 times or 62 % reduced, while the capacitance

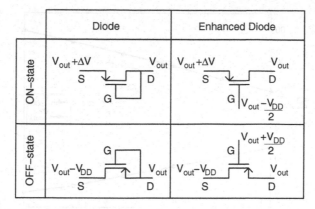

Fig. 6.20 Schematic view with bias levels of switch $S_{4,h}$ implemented as a diode or as a diode enhanced with a high-pass filter, both in its on-state (*forwardly biased*) and in its off-state (*reversely biased*)

in each Dickson stage is 4 pF compared to 2 pF. This actually doubles the expected output current of the converter while the output voltage is expected to be more or less the same. The equation below (6.11) estimates the efficiency η_3 of this design as a function of η_2, the efficiency of design 2.

$$\eta_3 \approx \frac{P_{out,3}}{P_{clk,3}} = \frac{2 \cdot P_{out,2}}{(1 - 0.62) \cdot P_{clk,2}} = 5.2 \cdot \eta_2 \tag{6.11}$$

where $P_{out,i}$ and $P_{clk,i}$ are respectively the output power and the power consumption for the clock circuitry in the converter design i. From this reasoning an efficiency increase with a factor 5.2 is expected for this design.

6.3.3.3 Measurement Results

The measured output voltage characteristics are presented in Fig. 6.21. The converter reaches output voltages of +47 V and −36 V for a power supply voltage of 20 V while the ring oscillator and the buffers consume 1.4 μA. The converter performs with a clock speed of 660 Hz. When a 10 nA output current is drawn the high output voltage $V_{out,h}$ lowers to 40 V. According to equation (6.11) the converter efficiency η_3 is 1.4 % for a 20 V V_{DD} and a 10 nA output current. The working point in which this efficiency is calculated is very similar to that used in 6.3.2.3 and the efficiency has increased by a factor of 1.5. The measurement results of the converter are summarized in Table 6.8. The chip photograph is shown in Fig. 6.22.

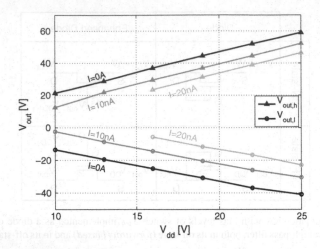

Fig. 6.21 The measured high and low output voltages versus the power supply voltage V_{DD} for a varying current from 0 to 20 nA

Table 6.8 Summary of the measurement results for the improved dual Dickson DC-DC converter integrated on a foil (Marien et al. 2011)

Specification	$V_{out,h}$	$V_{out,l}$
Power supply	20 V	
Clock frequency	660 Hz	
Voltage conversion ratio (@20 V)	2.5	2.5
Maximal output voltage (@20 V)	48 V	−33 V
Output current	20 nA	−20 nA
Output power	0.66 μW	
Tuning range	40 %	30 %
Oscillator and buffer current	1	μA
Chip area	$1.2 \times 1.5\,\text{mm}^2$	

6.4 Discussion

The measurements in Sect. 6.3 have revealed that the power efficiency of the presented converters is very low, below 5 %. This means that the converter itself consumes about 20 times more power than the power that it is able to deliver at the output. This remarkably poor result demands for a clarification.

The low efficiency is mostly provoked by the unipolar organic technology in which the circuits are applied. The highest power consumers in the circuit are the ring oscillator and the buffers. In the unipolar technology these are built with p-type transistors only for both the pull-up and the load (pull-down) transistor. This load transistor is implemented as a zero-V_{GS} transistor that behaves like a leaking current

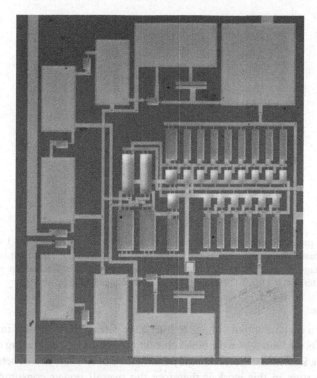

Fig. 6.22 Chip photograph of the improved dual Dickson DC-DC converter on foil. On this photograph the ring oscillator, the buffers and both Dickson cores are clearly distinguished. This circuit is made in technology provided by Polymer Vision

source hence the circuit consumes static power during half of the clock period as opposed to a complementary implementation where only dynamic power is consumed.

Moreover the buffers perform badly in pulling down the signals, since the pull-down component is a transistor biased in the cut-off region. Therefore these buffers are very large which results in an even higher power consumption. The design in Sect. 6.3.3 applies an alternative topology for the last buffer that improves the pull-down behavior and therefore improves the performance of the last buffer stage. However, this technique is not applicable in the ring oscillator because in every consecutive stage a voltage around V_T is lost preventing the ring from oscillating.

A third issue is related to the core efficiency that is calculated in Sect. 6.3.1.3. Since the diode-connected transistors perform far from ideally a not negligible current flows back in the converter at the output node and already degrades the optimal performance of the converter to about 50 % without even including the driving power of the clock and the buffers.

There are several reasons to believe that the efficiency of a converter is not the most appropriate figure of merit to judge about the organic DC-DC converters. The application field of organic DC-DC converters, discussed in Sect. 6.1, very clearly

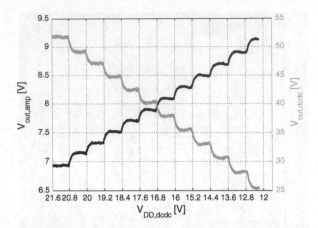

Fig. 6.23 The measured DC output voltage of the 2-stage op amp, from Fig. 3.2.1 shifts proportionally to the bias voltage generated by the integrated Dickson converter, from Fig. 6.19, of which the supply voltage is shifted. The gray curve corresponds to the measured output voltage of the DC-DC converter and the black curve to the DC output voltage of the op amp

demands for a certain output voltage to bias only gates and backgates in digital and analog circuits through which the direct current is negligible. The biasing of hundreds of gates or backgates is therefore already enabled by only a few nA. A better measure for the converters in this work is therefore the overall power consumption of the DC-DC converter. Moreover this power consumption must be low compared to the power applied in a certain circuit, e.g. the $\Delta\Sigma$ ADC in Sect. 4.3. The latter consumes $100\,\mu$A which is $10-100$ times more than the power consumed in any of the DC-DC converters presented in this dissertation. This measure of course only holds for this specific niche of the DC-DC converters sector but it supports the applicability of up-converters in organic electronics technology.

The proof of concept for the organic DC-DC converters is given in Fig. 6.23 where the measured DC output of the 2-stage op amp, discussed in Sect. 3.3.3, is shown with a bias voltage driven by the integrated DC-DC converter of which the supply voltage is varied. It is clearly seen that the DC level of the output of the op amp is at the desired DC level and that it is shifted in a linear way by the varying backgate bias voltage. With this measurement it is proven that the DC-DC converter can really drive other circuits and that it is a very useful block in organic electronics.

6.5 Conclusion

Designing both analog and digital organic circuits is an ongoing challenge. The research into processing techniques improving the technological environment and the single-transistor performance as well as the research into circuit techniques that enhance the behavior and the reliability of circuits is in motion. At the technology level, there is the ability to deposit a backgate on top of a transistor that influences its threshold voltage. At the circuit level, the backgate steering technique has been proven to improve the reliability of circuits. It is the combination of improvements at both research levels that demands for very high and very low bias voltages and enables the niche topic of organic DC-DC converters. The application field that demands for bias voltages rather than for power delivery at the outputs creates a whole different context for the DC-DC converters than the more common context present in standard *Si* technology. The atypical organic transistor technologies with their unusual and limited set of available components complicate the implementation of all kinds of circuits, also DC-DC converters. That is why a profound investigation is required into the feasibility of the implementation of several up-converter architectures in the organic electronics technologies.

In this chapter first an overview was given of the application field and three applications were highlighted: Digital circuit behavior is improved by biasing backgates to a voltage. Moreover analog circuits can be biased with their inputs and outputs at the same voltage level when a high bias voltage is available. Finally also a very low biasing voltage is beneficial for biasing transistors in the linear region. From these applications a set of specifications was deduced that are to be fulfilled in order to make organic DC-DC converters beneficial.

Next a broader comparison was provided between several types of DC-DC converters towards their applicability. Inductive converters are not applicable. At the moment the Dickson converter architecture is the best fit for application in organic technology. It scores better than series-parallel, voltage doubler and Fibonacci architectures for complexity and switch implementation.

Subsequently three Dickson converter designs were presented with their implementation, simulation results and measurement results.

Finally a discussion about the very low power efficiency of the presented converters was provided. This low power efficiency was mostly caused by the limitations of the unipolar circuit technology. Because of their distinct application field, however, organic DC-DC converters were considered an added value to organic circuit design.

References

Myny K, Beenhakkers MJ, van Aerle NAJM, Gelinck GH, Genoe J, Dehaene W, Heremans P (2011) Unipolar organic transistor circuits made robust by Dual-Gate technology. IEEE J Solid-State Circ 46(5):1223–1230

Van Breussegem T, Steyaert M (2010) A fully integrated 74% efficiency 3.6 V to 1.5 V 150 mW capacitive point-of-load DC/DC-converter. In: Proceedings of the ESSCIRC 2010. pp 434–437

Wens M, Cornelissens K, Steyaert M (2007) A fully-integrated 0.18m CMOS DC-DC step-up converter, using a bondwire spiral inductor. In: 33rd European Solid State Circuits Conference ESSCIRC 2007. pp 268–271

Wens M, Steyaert MSJ (2011) A fully integrated CMOS 800-mW Four-Phase semiconstant ON/OFF-Time Step-Down converter. IEEE Trans Power Electron 26(2):326–333

Dickson JF (1976) On-chip high-voltage generation in MNOS integrated circuits using an improved voltage multiplier technique. IEEE J Solid-State Circ 11(3):374–378

Marien H, Steyaert M, Steudel S, Vicca P, Smout S, Gelinck G, Heremans P (2010) An organic integrated capacitive DC-DC up-converter. In: Proceedings of the ESSCIRC 2010. pp 510–513

Marien H, Steyaert M, van Veenendaal E, Heremans P (2011) Organic dual DC-DC upconverter on foil for improved circuit reliability. Electronics Lett 47(4):278–280

Chapter 7
Conclusions

7.1 General Conclusions

In this dissertation the feasibility of analog and mixed-signal circuits in organic thin-film transistor technology on foil has been investigated and demonstrated. Multiple building blocks, connected through the backbone of an umbrella application, i.e. organic smart sensor systems, have been considered. This book is subdivided in three parts:

- an introduction to the domain and the application field of organic electronics,
- a discussion of the existing organic electronics technology on foil with its main advantages and disadvantages,
- the presentation of multiple·analog and mixed-signal building blocks which are employed in a proposed organic smart sensor system architecture.

The first part coincides with Chap. 1 and discusses the important properties and the application field of organic electronics. The main properties of organic electronics are the low-cost and flexible plastic substrate and its printability, which all follow from the low processing temperatures (150 °C). This technology is applicable in flexible, large-area and large-scale application, e.g. for flexible displays, for RFID applications and for organic smart sensor systems. For the analog interface between the sensor and the digital part of this last application a system architecture is proposed.

The second part of this work is a deeper exploration of the existing organic electronics technology. This part is treated in Chap. 2. After an overview of the available materials and the existing deposition techniques, the behavior of the bottom-gate pentacene thin-film transistors in the unipolar technology is investigated. The 3-contact architecture of this transistor has a MOSFET-like behavior. A 4-contact architecture is discussed with an additional backgate. This backgate contact has a linear influence on the V_T of the transistor and it turns out that it is a very useful contact for overcoming non-ideal behavior of the transistors. Such non-ideal behavior is observed when the transistors are biased in the presence of ambient environment, light and others, and often results in a change of behavioral parameters V_T and μ. Besides

H. Marien et al., *Analog Organic Electronics*, Analog Circuits and Signal Processing, 161
DOI: 10.1007/978-1-4614-3421-4_7, © Springer Science+Business Media New York 2013

the integrated transistors, also the possibility to integrate passive components in the technology is explored.

The third and largest part of this research work is the implementation and the demonstration of multiple analog and mixed-signal building blocks. This part is covered by Chaps. 3–6, which each discuss the implementation and the measurement results of one building block of the proposed organic smart sensor system architecture.

The differential amplifier, in Chap. 3, is the first analog building block that is investigated. First a defensive design methodology is applied to a single-stage differential amplifier, focusing on both high gain and high reliability of the amplifier. Simulation results regarding the reliability and measurement results of the performance of this amplifier are presented. Furthermore, the design and the measurements of a 3-stage opamp is presented, which employs high-pass filters as a throughput between the consecutive stages. In order to overcome those high-pass filters, an improved architecture is presented of a single-stage amplifier that enables the direct connection of consecutive stages and measurements of both the single-stage amplifier and a 2-stage DC connected opamp are presented.

The knowledge about amplifier design is applied to a more complex building block, i.e. an ADC, in Chap. 4. First a study of the possible ADC architecture families is carried out and their feasibility in organic electronics technology is examined. As a conclusion of this study the $\Delta\Sigma$ ADC is preferred over other architectures for its low variability and for the flexible trade-off between speed and accuracy. Then the implementations of a $1st$-order and a $2nd$-order $\Delta\Sigma$ ADC on foil are elucidated and a measured accuracy of 26.5 dB is presented.

The sensor is the building block of the smart sensor system that determines the core function of the system. Integrated sensors are the subject of Chap. 5. The possible sensors and their application field are discussed first. Subsequently the implementation of both a 1D and a 2D 4 × 4 touch pad are presented. Both are capacitive sensors which are based on the series connection of capacitors created when a finger is located on top of the sensor. The variable capacitance is measured through a current mirror and a known output capacitor. The sensor read-out has a measured sample rate of 1.5 kS/s. According to simulations an interpolated accuracy is obtained of 0.3 mm in a total range of 35 mm.

The most atypical building block of the proposed organic smart sensor system is the DC–DC up-converter, presented in Chap. 6. This building block is the answer to a direct need for bias voltages, higher than the power supply voltage and lower than the ground voltage, in the implementations of the other building blocks discussed in this work. In a profound survey the optimal up-converter architecture is looked for and it has turned out that a Dickson converter has the lowest complexity implementation in organic electronics technology. Three designs of integrated Dickson converters are included in this chapter and their measurement results have been presented. Finally, it has been demonstrated that the implemented converter is able to drive the desired bias pin of the 2-stage opamp with the desired voltage of 35 V.

Index

H. Marien et al., *Analog Organic Electronics*, Analog Circuits and Signal Processing, 163
DOI: 10.1007/978-1-4614-3421-4, © Springer Science+Business Media New York 2013